# Tissue Engineering
# for Degenerative
# Intervertebral Discs

## Other Related Titles from World Scientific

*Stem Cell and Tissue Engineering*
edited by Song Li, Nicolas L'Heureux and Jennifer Elisseeff
ISBN: 978-981-4317-05-4

*Stem Cells, Tissue Engineering and Regenerative Medicine*
edited by David Warburton
ISBN: 978-981-4612-77-7

*Stem Cell Bioengineering and Tissue Engineering Microenvironment*
edited by Satya Prakash and Dominique Shum-Tim
ISBN: 978-981-283-788-2

*Tissue Engineering and Nanotheranostics*
edited by Donglu Shi and Qing Liu
ISBN: 978-981-3149-18-2

# Introduction

Low back pain is a chronic disease that plagues the departments of orthopedics owing to its high incidence and choice of proper clinical treatments. Lumbar intervertebral disc degeneration is one of the main reasons for a visit to the orthopedics department. Lumbar intervertebral disc degeneration is difficult to reverse. Traditional treatment methods include conservative treatment and surgical treatment. Although the clinical symptoms caused by intervertebral disc degeneration can be alleviated to a certain extent, it cannot solve the fundamental problem, and it also produces other complications. In addition, nucleus pulposus removal from the intervertebral disc makes the disc lose its basic function, and so joint activities will inevitably accelerate the degeneration of adjacent intervertebral discs.

The ideal surgical treatment is repair. The human intervertebral disc is made up of three parts: annulus fibrosus, nucleus pulposus, and cartilage endplate. It is flexible and strong, and not only functions to maintain the stability of the spine, transfer loads, absorb shocks and disperse stress, but also forms the basis of the structure that is a major player in the flexibility of spinal movement. The integrity of the intervertebral disc is of great significance for maintaining the normal physiological function of the spine. However, in clinical work, often due to various reasons, the nucleus pulposus has to be removed, such as in serious cases of lumbar disc herniation, to ease the clinical symptoms. Experimental results show that after discectomy, the intervertebral space will narrow, the anterior lumbar stress will decrease, and back stress will increase, resulting in spinal disorders relating to biomechanical functioning, lumbar instability, and accelerated degeneration of adjacent segments. Because of the complexity of the biomechanical structure of the intervertebral disc, it is difficult to replace it with any kind of artificial intervertebral disc, as in most cases the long-term effects of artificial intervertebral discs are not satisfactory. In recent years,

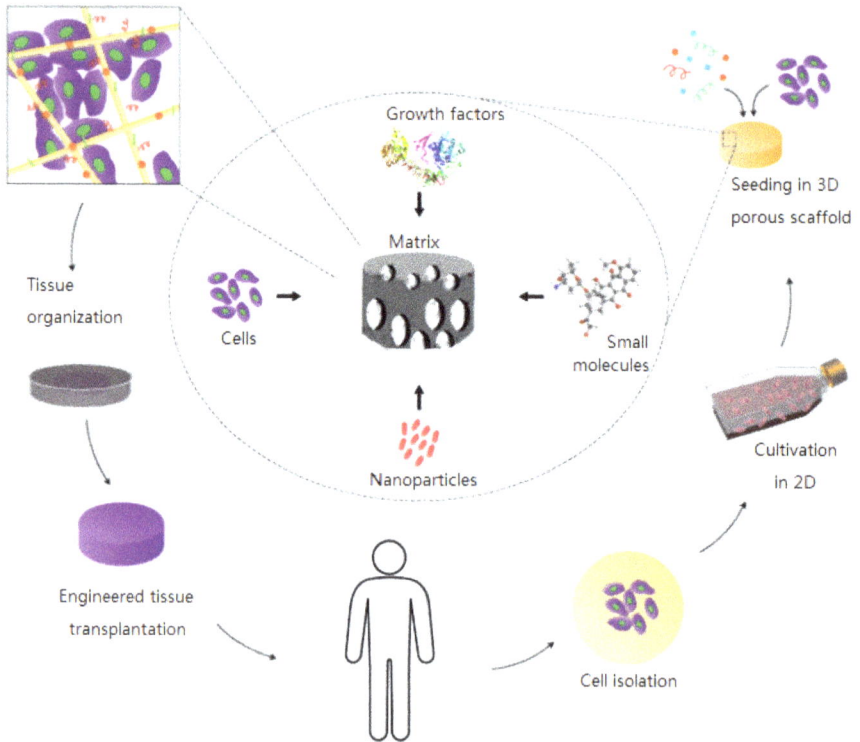

Fig. 1. The rise of tissue engineering technology and its applications in different fields have brought about new ideas for the treatment of intervertebral disc degeneration. Reprinted with Creative Commons Attribution License (CC BY 4.0) from Figure 1 in Ref. [1].

with the rapid development of tissue engineering and its successful application in the repair of bone and articular cartilage defects, scientists have begun to try and use the method of tissue engineering reconstruction for damaged discs, and some achievements have been made.

Tissue engineering is based mainly on cells, including seed cells, which are then made to attach to the biological scaffold or extracellular matrix, which is used to promote tissue regeneration. The rise of tissue engineering technology and its application in different fields have brought about new ideas for the treatment of intervertebral disc degeneration (Fig. 1).

## Reference

1. Courtenay JC, Sharma RI, Scott JL. Recent advances in modified cellulose for tissue culture applications. *Molecules*, 2018, 23(3): 654.

# Chapter 1

# Progress in Tissue Engineering

Various types of tissue and organ dysfunctions contribute to human health issues, eventually leading to death. Tissue defect repair and functional reconstruction are some of the key challenges in the medical field. Traditional treatments often require sacrifice of autologous tissue for transplantation and repair, which can easily lead to trauma and function loss of the donor area. It is futile to repair trauma with trauma, and allograft organ transplantation is greatly limited because of lack of suitable donors. With the development of medical science, the concept of tissue and organ defects has gradually changed from tissue transplantation to tissue regeneration. Tissue engineering, as an important component of regenerative medicine and an important means of tissue regeneration, has developed rapidly in the past 20 years and more [1–3]. If we get a small amount of autologous tissue from the body, extract seed cells for *in vitro* cultivation on scaffold materials, through the formation of engineered tissues and after implantation *in vivo*, repair tissue defect and restore the original function of tissue, then this mode can avoid the defects of wound healing and is expected to achieve tissue regeneration and functional reconstruction using noninvasive or minimally invasive procedures.

In this text, important research results are reviewed: the application of stem cells, development from a simple structure of the scaffold material to the intelligent material structure that is complex, with biological activity, simulation of local micro-environment, with the immune function of animal body for the repair of bone, cartilage, and tendon. The clinical studies of tissue engineering have been carried out in many tissue defects, such as skin and other tissue defects. These research results have laid the foundation for further development. The following is a brief introduction to the main progress made in recent years in the repair of bone, cartilage, tendon and blood vessels [4, 5].

## 1. Tissue Engineering of Bone

With the advancements in adult stem cell tissue engineering, bone construction technology and defect repair, it is now well known that the clinical bone defect repair caused by trauma, tumor and congenital malformations has been an important issue in the field of plastic surgery and department of orthopedics. Compared with traditional treatments, the application of tissue-engineered bone to repair bone defect not only can help avoid the problem of limited, existing treatments of autogenous bone graft by bone graft donor, but also overcome the lack of bone substitute biomaterials, without the limitation of physiological repair. The basic and clinical application of tissue-engineered bone has been carried out since early on. In recent years, bone marrow mesenchymal stem cells (BMSCs) have been used to construct tissue-engineered bone. Animal experimental studies on the repair of femoral defects, tibia defect, defect of mandible, skull defect, knee joint cartilage defect, bone defect and other types of bone defects are carried out. For the clinical application of bone tissue engineering, the patient's autologous BMSCs are used as seed cells. Tissue engineering technology has been used to repair skull defect, alveolar cleft defect, inferior turbinate defect, facial bone deformity and limb bone defect. The longest follow-up time was 5 years [6]. The results showed that tissue-engineered bone could be stable for a long time in the patients. It could basically restore the appearance of the bone defect area, and also carry out the functions of normal bone support and protection. All of the above results show that using autologous BMSCs as seed cells, tissue engineering technology can be used to form stable tissue-engineered bone tissue and help repair bone defects in the human body, thereby laying a solid foundation for large-scale clinical application of tissue-engineered bone [7–9].

## 2. Tissue Engineering of Cartilage

In order to solve the problem of the limited source of seed cells in bone tissue engineering research, the adipose-derived stem cells (ADSCs) were channeled into a series of osteogenic differentiation [10]. The main discoveries include the role of the ERK pathway in osteogenic differentiation and in the regulation of ADSCs; understanding of the molecular mechanism of bone ADSCs dexamethasone interference differentiation, immune ADSCs inhibition and elucidation of its mechanism; application of allogeneic ADSCs for the repair of a canine skull defect; and demonstration of the good osteogenic activity after *in vitro* cryopreservation of ADSCs, which provided the basis for the establishment of an allogeneic stem cell bank.

Since cartilage construction technology is based on adult stem cells and cartilage tissue regeneration ability is extremely low, cartilage defect or injury has always been a difficult problem in surgical treatment. In order to solve this problem, an ideal treatment method is provided [11–15]. However, due to the limited source of cartilage cells and the large area of trauma, the research into and application of tissue-engineered cartilage based on chondrocytes have been greatly limited. The discovery of adult stem cells and the progress made at present have brought hope to solve the problem of cartilage seed cells.

However, most of the international studies are still limited to the study of cartilage-induced differentiation in stem cells and the construction of cartilage based on stem cells. There has been no breakthrough in the technology of repair of large animal cartilage defects, which seriously hinders the transfer of this technique into clinical application. In recent years, there has been some progress in stem cell research and the advancement of cartilage repair technology of large animal cartilage defects, *in vitro* cartilage construction, and repair related to swine autologous articular cartilage and bone defects by the early application of BMSC composite. One of the breakthroughs was the microenvironment for the construction of cartilage, in which BMSCs and chondrocytes were cultured and the environment had successfully constructed mature cartilage. Thus, through a series of co-culture models, the key mechanism of BMSC-induced differentiation of chondrocytes in cartilage [16–21] was explained. The knowledge on construction technology of *in vitro* cartilage, establishment of the core technology of the application of BMSCs in *in vitro* chondrogenesis, and the evaluation of *in vitro* chondrogenesis *in vivo*, especially through computer-aided support design and preparation, for the first time helped to solve the human auricular cartilage reconstruction problem and form the process of constructing an accurate model. With the reconstructed cartilage, which had the exact appearance of auricle, *in vitro* studies and studies on mice were successfully performed.

BMSCs were used for the repair of autologous knee joint cartilage and bone defect of large area. Through the surgical technique of pedicle grafting to improve long-term survival, a rabbit tracheal defect was repaired for the first time using autologous tissue engineering; in addition, autologous fibroblasts were also used, for the first time, for the repair of canine meniscus defect. These advances and breakthroughs have solved the core problem of cartilage tissue engineering in many ways and greatly promoted the transfer of cartilage tissue engineering from research to clinical practice.

## 3. Tissue Engineering of Tendons

The basic research and application of tissue engineered tendons is based on the study of biomechanics in tissue engineering. Based on work that has been carried out, the following roles and mechanisms in tendon construction have been identified: (1) The cells show a narrow state, and one-way mechanical stimulation can help to maintain the phenotype of tendon cells *in vitro*; in these biological processes, the Rho A/Rock signal transduction system plays a key role in hypoxia; and proliferation is a function that is helpful for the maintenance of the tendon cells [22–24]. (2) The narrow morphology and unidirectional mechanical stimulation can promote the trans-differentiation of the skin fibroblasts to tendon cells, which thus become a new source of seed cells. (3) Tissue-engineered tendons are constructed *in vitro* by human skin fibroblasts, and mechanical stimulation can promote their maturation.

Under mechanical stimulation, ADSCs could form tendon and repair the rabbit Achilles tendon defect. Using the chicken model construct, tissue-engineered sheath tissue with certain functions was created [25–28].

Relevant basic research has been carried out in collaboration between the University of Pittsburgh in USA and the Kyoto University in Japan. In the field of applied research, the work is focused on the early clinical study of tendon repair in tissue-engineered tendons, and the following four aspects have been studied:

(1) A mechanical reinforced tendon scaffold material has been developed, which not only has good cell/biocompatibility, but also can form a tissue-engineered tendon with mechanical strength up to 50 N *in vitro*.

(2) The use of mechanical reinforced scaffold materials and skin fibroblasts *in vitro* to construct tendons. The repair of flexor tendon defect in the two areas of a monkey's hand was successfully done by implanting the constructed tendons in the macaque, and scientific basis was established for clinical application.

(3) The safety of using skin fibroblasts as seed cells was assessed by carrying out chromosome stability assays, tumor formation experiments, cell and matrix secretion assays, and allergy and pathogen detection, all of which not only confirmed the amplification of human fibroblasts *in vitro* with tendon formation but also the safety. Thus, this technique has received a certificate from the National Institute for the Control of Pharmaceutical and Biological Products, for providing security and protection for the development of muscle tendon products.

(4) The localization and production of tendon scaffold materials have been carried out. Construction and development of the production equipment, which can reduce the price of the scaffold fiber material, have also been set up in a corresponding factory.

## 4. Tissue Engineering of Vascular Structures

Cardiovascular disease has become the number one killer in modern society. Vascular transplantation is currently a relatively effective treatment. The application of tissue engineering technology to construct blood vessels and enable the repair of vascular defects *in vivo* has confirmed the feasibility of clinical application of tissue-engineered blood vessels, which is expected to provide new ideas for solving the problem of vascular graft donors with biological function [29–31].

However, this technique has not been transferred into practical clinical application because of the limitations of cell origin and the poor stability of vascular construction technology. Studies on vascular seed cells and construction technology in recent years have made significant progress. The main research results are as follows. A tissue-engineered blood vessel perfusion pulsatile bioreactor system was successfully constructed with the biomechanical strength of the vascular lumen using a bioreactor, a preliminary solution in the construction of vascular grafts with biomechanical properties to solve the basic problems. In 2008, the introduction of a new type of bioreactor from the American company Bose was used to construct blood vessels for circulation and perfusion, to apply a pulse stimulation, to enable automatic control over flow rate, pulse frequency and duration of all relevant parameters and hemodynamics to simulate normal blood vessels. The seed cells were successfully established: purification, amplification and differentiation induced orientation of endothelial progenitor cells in the bone marrow, and a new method for obtaining a large amount of vascular endothelial progenitor cells was found in the process of high-density culture of bone marrow cells, using vascular endothelial cells to expand the source. Through the application of a variety of growth factors, BMSCs and ADSCs are successfully directed into cells of smooth muscle phenotype and function, which extend the source of vascular endothelial cells. With regard to technical aspects, BMSCs and ADSCs have successfully been used to construct the vascular wall with good elasticity under the mechanical stimulation of a bioreactor. These advances and breakthroughs have solved the core issue of vascular tissue engineering from many aspects and have made positive contributions to promoting the transformation of vascular tissue engineering from basic research to clinical application.

# 5. Alternative Sources of Seed Cells

Seed cells are the foundation of tissue-engineered reconstruction of tissues. Stem cells have become the focus of seed cell research because of their powerful amplification and multipotential differentiation capabilities. Mesenchymal cells derived from the bone marrow have been successfully applied to the repair of bone and cartilage defects because of their osteogenic and chondrogenic differentiation. However, bone marrow injury is relatively large, and the acceptance by patients is low. Also, whether a large number of extracts will affect hematopoietic function is currently not clear. Therefore, searching for other tissue-derived pluripotent stem cells has become the focus of research. The skin is an ideal source of cells owing to its large surface area enabling the convenient selection of material. Studies have shown that the epidermal layer contains epidermal stem cells, hair follicle stem cells and other pluripotent cells, but whether there are stem cells in the dermal dermis is still uncertain. Since normal skin also undergoes periodic renewal and replacement, we speculate that the dermis also contains multiple differentiated stem cells.

# References

1. Zhu L, Liu W, Cui L, *et al*. Tissue-engineered bone repair of goat femur defects with osteogenically induced bone marrow stromal cells. *Tissue Eng*, 2006, 12(3): 423–433.
2. Weng Y, Wang M, Liu W, *et al*. Repair of experimental alveolar bone defects by tissue-engineered bone. *Tissue Eng*, 2006, 12(6): 1503–1513.
3. Yuan J, Cui L, Zhang WJ, *et al*. Repair of canine mandibular bone defects with bone marrow stromal cells and porous beta-tricalcium phosphate. *Biomaterials*, 2007, 28(6): 1005–1013.
4. Liu G, Zhao L, Zhang W, *et al*. Repair of goat tibial defects with bone marrow stromal cells and $\beta$-tricalcium phosphate. *J Mater Sci: Mater Med*, 2008, 19(6): 2367–2376.

5. Yuan J, Zhang WJ, Liu G, *et al.* Repair of canine mandibular bone defects with bone marrow stromal cells and coral. *Tissue Eng Part A*, 2010, 16(4): 1385–1394.

6. Liu W, Cui L, Cao Y. Bone reconstruction with bone marrow stromal cells. *Methods Enzymol*, 2006, 420: 362–380.

7. Cui L, Yin S, Liu W, *et al.* Expanded adipose-derived stem cells suppress mixed lymphocyte reaction by secretion of prostaglandin E2. *Tissue Eng*, 2007, 13(6): 1185–1195.

8. Liu Q, Cen L, Zhou H, *et al.* The role of the extracellular signal related kinase signaling pathway in osteogenic differentiation of human adipose-derived stem cells and in adipogenic transition initiated by dexamethasone. *Tissue Eng Part A*, 2009, 15(11): 3487–3497.

9. Liu X, Sun H, Yan D, *et al. In vivo* ectopic chondrogenesis of BMSCs directed by mature chondrocytes. *Biomaterials*, 2010, 31(36): 9406–9414.

10. Liu K, Zhou GD, Liu W, *et al.* The dependence of *in vivo* stable ectopic chondrogenesis by human mesenchymal stem cells on chondrogenic differentiation *in vitro*. *Biomaterials*, 2008, 29(14): 2183–2192.

11. Yan D, Zhou G, Zhou X, *et al.* The impact of low levels of collagen IX and pyridinoline on the mechanical properties of *in vitro* engineered cartilage. *Biomaterials*, 2009, 30(5): 814–821.

12. Liu Y, Zhang L, Zhou G, *et al. In vitro* engineering of human ear-shaped cartilage assisted with CAD/CAM technology. *Biomaterials*, 2010, 31(8): 2176–2183.

13. Li Q, Liu T, Zhang L, *et al.* The role of bFGF in down-regulating $\alpha$-SMA expression of chondrogenically induced BMSCs and preventing the shrinkage of BMSC engineered cartilage. *Biomaterials*, 2011, 32(21): 4773–4781.

14. Zhou G, Liu W, Cui L, *et al.* Repair of porcine articular osteochondral defects in non-weightbearing areas with autologous bone marrow stromal cells. *Tissue Eng*, 2006, 12(11): 3209–3221.

15. Jiang T, Liu W, Lv X, *et al.* Potent *in vitro* chondrogenesis of CD105 enriched human adipose-derived stem cells. *Biomaterials*, 2010, 31(13): 3564–3571.

16. Zhu J, Li J, Wang B, *et al.* The regulation of phenotype of cultured tenocytes by microgrooved surface structure. *Biomaterials*, 2010, 31(27): 6952–6958.

17. Jiang Y, Liu H, Li H, *et al.* A proteomic analysis of engineered tendon formation under dynamic mechanical loading *in vitro*. *Biomaterials*, 2011, 32(17): 4085–4095.

18. Zhang Y, Wang B, Zhang WJ, *et al.* Enhanced proliferation capacity of porcine tenocytes in low O2 tension culture. *Biotechnol Lett*, 2010, 32(2): 181–187.

19. Deng D, Liu W, Xu F, *et al.* Engineering human neo-tendon tissue *in vitro* with human dermal fibroblasts under static mechanical strain. *Biomaterials*, 2009, 30(35): 6724–6730.

20. Cao D, Liu W, Wei X, *et al. In vitro* tendon engineering with avian tenocytes and polyglycolic acids: A preliminary report. *Tissue Eng*, 2006, 12(5): 1369–1377.

21. Liu W, Chen B, Deng D, *et al.* Repair of tendon defect with dermal fibroblast engineered tendon in a porcine model. *Tissue Eng*, 2006, 12(4): 775–788.

22. Xu L, Cao D, Liu W, *et al. In vivo* engineering of a functional tendon sheath in a hen model. *Biomaterials*, 2010, 31(14): 3894–3902.

23. Xu ZC, Zhang WJ, Li H, *et al.* Engineering of an elastic large muscular vessel wall with pulsatile stimulation in bioreactor. *Biomaterials*, 2008, 29(10): 1464–1472.

24. Wang C, Cen L, Yin S, *et al.* A small diameter elastic blood vessel wall prepared under pulsatile conditions from polyglycolic acid mesh and smooth muscle cells differentiated from adipose-derived stem cells. *Biomaterials*, 2010, 31(4): 621–630.

25. Wang C, Yin S, Cen L, *et al.* Differentiation of adipose-derived stem cells into contractile smooth muscle cells induced by transforming growth factor-beta1 and bone morphogenetic protein-4. *Tissue Eng Part A*, 2010, 16(4): 1201–1213.

26. Chen FG, Zhang WJ, Bi D, *et al.* Clonal analysis of nestin(−) vimentin(+) multipotent fibroblasts isolated from human dermis. *J Cell Sci*, 2007, 120(Pt 16): 2875–2883.

27. Yin S, Cen L, Wang C, *et al.* Chondrogenic trans differentiation of human dermal fibroblasts stimulated with cartilage-derived morphogenetic protein 1. *Tissue Eng Part A*, 2010, 16(5): 1633–1643.

28. Zhao G, Yin S, Liu G, *et al. In vitro* engineering of fibrocartilage using CDMP1 induced dermal fibroblasts and polyglycolide. *Biomaterials*, 2009, 30(19): 3241–3250.

29. Liu W, Chen B, Deng D, *et al.* Repair of tendon defect with dermal fibroblast engineered tendon in a porcine model. *Tissue Eng*, 2006, 12(4): 775–788.

30. Zhang YQ, Zhang WJ, Liu W, *et al*. Tissue engineering of corneal stromal layer with dermal fibroblasts: Phenotypic and functional switch of differentiated cells in cornea. *Tissue Eng Part A*, 2008, 14(2): 295–303.
31. Bi D, Chen FG, Zhang WJ, *et al*. Differentiation of human multipotent dermal fibroblasts into islet-like cell clusters. *BMC Cell Biol*, 2010, 11: 46.

# Chapter 2

# Stem Cells and Regenerative Medicine Research

Stem cells are a group of cells that are capable of self-renewal and multidifferentiation and include embryonic stem cells isolated from embryos, induced pluripotent stem cells, and adult stem cells. Stem cell is an important research subject to help in the understanding of the basic mechanisms of the pluripotent maintenance, somatic cell mechanism, and disease pathogenesis. Stem cells have also been known to play an important role in the treatment of hereditary diseases and the construction of "seed cells" *in vitro*, which are of great value in the treatment of diseases and in the field of regenerative medicine. In recent years, many Chinese scientists in the field of stem cells have made great progress, especially in research concerning induced pluripotent stem cells and reprogramming, trans-differentiation, haploid stem cells and adult stem cells and biological materials. This field of study combined with a genetically modified animal model study and gene therapy is particularly prominent.

## 1. Induction of Pluripotent Stem Cells

In 2006, the Yamanaka Nobuya research group first discovered that 4 transcription factors, Oct4, Sox2, KLF4, and c-Myc, can be used to reprogram somatic cells into pluripotent stem cells, that is, the induced pluripotent stem cells (iPSCs) [1]. Induced pluripotent stem cells have effectively solved the ethical controversy and the immunological rejection problems seen with embryonic stem cells (ESCs) and have opened a new door for stem cell research. However, iPSCs have not been able to compensate for the development of tetraploid cells like embryonic stem cells. Therefore, whether iPSCs can develop into different kinds of cells *in vivo* by using the gold-standard pluripotent stem cells is a challenging problem in the field. In order to answer this question, Chinese scientists obtained [2, 3] iPSCs from mice and their offspring successfully through the establishment of newly induced culture systems and tetraploid compensation technology. This work has fully demonstrated the developmental omnipotency of iPSCs, which has paved the way for the clinical and basic research on iPSCs. However, not all iPSCs have the complete potential for development. Only a small portion of them can undergo tetraploid compensation. The question thus arises: Are there any differences in iPSCs with different developmental potentials? Chinese scientists compared different developmental potentials of iPSCs at the molecular level, for the first time, and found that abnormal expression of the imprinted gene Dlk1–Dio3 led to the lowering of the ability of iPSCs. They also found that the developmental ability of Dlk1–Dio3 cells in the tetraploid mark area is normal [4]. For the first time, this work identified a marker for differentiating iPSCs of different developmental potentials. It is of great reference value for us to study the developmental potential of human pluripotent stem cells, and this has become a recognized standard in the international stem cell field. iPSCs induced by low efficiency have been an important factor limiting the application of iPSCs, but Chinese scientists found that vitamin C can significantly improve the induction efficiency of iPSCs. Through a series of studies, it was revealed how vitamin C helps modify the epigenetic status to influence reprogramming [5–7]. It was found that vitamin C not only improved the efficiency of reprogramming but also had a significant impact on the quality of the reprogramming of the obtained iPSCs. Thus, adding vitamin C can help to effectively improve the development ability of iPSCs and help to maintain the stability of genes [8]. Thus, vitamin C is extremely important to improve the quality and efficiency of reprogramming. In addition to vitamin C, many small molecules that are capable of improving the efficiency of reprogramming of transcription factors have been found; these small molecules can replace some reprogramming to a certain extent, and the use of small molecules to reduce or completely replace the transcription factors not only can improve the security of iPSCs but is also of great value to improve the standard applications of iPSCs in the future. Although many small molecules can effectively improve the efficiency of reprogramming and can replace some transcription factors, until 2013 Chinese scientists, using only a combination of 7 small molecules, successfully altered the transcription factors in mouse somatic cells, thereby enabling their conversion into pluripotent stem cells, which have higher development ability and the capability to form different types of cells [9]. Meanwhile, in the reprogramming mechanism research, we found that transcription factors related to differentiation can replace pluripotent transcription factors to induce somatic cell reprogramming, and so we propose a molecular regulation model [10], where transformation occurs from adult cells to pluripotent stem cells. Scientists have discovered many mechanisms of reprogramming based on induced pluripotent stem cells, and have further developed this technology on the basis of previous work, which promoted the technology to enter the clinic as quickly as possible.

## 2. Trans-differentiation

Somatic cells under the influence of certain transcription factors can be reprogrammed into another type of cells; this is called trans-differentiation. Trans-differentiation without a pluripotent stem cell stage and a relatively rich medium enables one to easily get the cell into a state with relative lack of its important functions, thus reducing the possibility of cancer. Trans-differentiation can occur within a relatively short time, and can thus answer the urgent need of cell transplants as the new choice. Early studies of trans-differentiation originated in 2002, and scientists used MyoD to convert fibroblasts into myoblasts [11]. Although trans-differentiation occurs faster, many stem cells appeared before the development of this technology by using the process of slow induction, that is, following the emergence of iPSCs and differentiation. This became the focus of everyone's attention, and since then various types of trans-differentiation studies, similar to the bamboo shoots that appear after a spring rain, have been performed. They first established the technical system of neural stem cell trans-differentiation, by successfully inducing mesoderm Sertoli cells to differentiate into neural stem cells, and then these neural stem cells can differentiate into various types of neural cells [12] *in vitro* and *in vivo*. Neural stem cells have the ability to proliferate and have a multi-directional differentiation potential to create neurons; therefore, they have more important value in terms of clinical applications. Chinese scientists also took the lead in inducing mouse fibroblasts to differentiate into liver cells, and more importantly, the liver cells were transplanted into the liver of mice with liver disease and it was shown that they could save the lives of mice [13]. This indicates that the trans-differentiated liver cells obtained possess all or part of the function after transplantation, thus providing an important reference for the future clinical application of trans-differentiation.

The same is true for the differentiation of hepatocytes. In the follow-up study, Chinese scientists have successfully obtained hepatic precursor cells by trans-differentiation, as well as human hepatocyte-like cells [14–16]. However, trans-differentiation and production of induced pluripotent stem cells with the help of exogenous transcription factors to achieve cell fate conversion and the introduction of exogenous genes to promote trans-differentiation pose a potential threat. In order to solve this problem, our scientists combined 3 small molecular compounds and created a hypoxic environment and successfully converted mice and human somatic cells into neural precursor cells [17], which can effectively improve the safety of cells obtained by trans-differentiation.

## 3. Adult Stem Cells and Biomaterials

Adult stem cells are those that exist in already differentiated tissues but have self-renewal ability and pluripotency. Compared with embryonic stem cells, these cells have many sources, low immune rejection, and low oncogenic risk, are less controversial, and are more ethical, besides having other advantages, thus signaling great hope for regenerative medicine. In the field of cell therapy, adult stem cells have good application prospects, especially the use of mature adult stem cells including hematopoietic stem cells and mesenchymal stem cells (MSCs), due to the limited expansion capacity of hematopoietic stem cells *in vitro*. In recent years, MSCs have gained more attention in basic research and for clinical applications. MSCs exist in a variety of tissues and play an important role in the repair of tissue damage [18]. Although MSCs do not have high differentiation potential as do embryonic stem cells, MSCs have their own unique advantages. First, MSCs have no ethical constraints regarding use for autologous or allogeneic transplantation; second, MSCs have a wide range of sources, and based on the body damage, researchers have isolated MSCs from various tissues [19]; third, MSCs have a strong autocrine and paracrine function and can secrete bioactive substances to provide nutritional support and aid in immune regulation of tissue repair [20]. These advantages have laid a foundation for the clinical application of MSCs. Therefore, it is necessary to carry out, in depth, a comprehensive and basic research on MSCs and to study the animal model of the disease before these are utilized in the clinic. The question of whether stem cells can cause immune rejection is an important issue. Our scientists have systematically studied the interaction between MSCs and the immune system and have elucidated the effects of MSCs on the immune system.

This has paved the way for the clinical application of MSCs, thus opening new doors for further research in this field [21, 22]. After this, MSCs led to some breakthroughs in clinical application, as scientists have been carrying out experiments using MSCs for the treatment of graft-versus-host disease (GVHD), aplastic anemia (AA), arthritis, systemic lupus erythematosus, and other autoimmune diseases [23, 24]. In addition to autoimmune diseases, some studies have also shown that MSCs can differentiate into germ layer cells and possess the ability to differentiate into the endoderm and ectoderm. Chinese scientists have used MSCs to treat spinal cord injury [25] (http://clinicaltrials.gov/) and other conditions such as epithelial injury, pulmonary fibrosis, cerebral palsy, Alzheimer's disease, diabetes, cardiovascular disease, liver diseases, burns, and nerve injury. Some clinical trials using MSCs have been carried out, such as the treatment of type I diabetes with MSCs, which are currently in phase I and stage II.

There are some problems in cell transplantation, such as reduction, loss, and diffusion of cell activity after a single cell transplantation, thus making it difficult to achieve the desired therapeutic effect. With the development of biological materials,

Chinese scientists first used small molecules (such as bFGF, VEGF) in combination with biological materials in animal disease models [26, 27] and achieved good treatment effects; by doing so, the Chinese scientists found that the MSCs cultured on 3D scaffolds had better pluripotency [28, 29] than cells cultured on 2D scaffolds. Thus, after the success of a combination of biological materials and mouse embryonic stem cells in culture, combinations of different stem cells and biological materials were tried. For example, the Institute of Genetics and Development of the Chinese Academy of Sciences combined MSCs and collagen in a rat model of brain injury and found that these cells had a better effect on tissue repair after binding [30]. However, it must be noted that when cells are combined with biological materials, they will be faced with a new environment. The properties of the materials will have various effects on cells, such as changes in pH, conductivity, pressure, and some other stimulative effects [31]. Therefore, finding a biological material that is as close to and safe for the physiological environment is the required direction to be taken in the field of biomaterials.

## 4. Haploid Stem Cells

Haploid cells have only a single set of chromosomes, so they are of great value in genetic screening and gene function research. Haploids are commonly seen in lower biological bacteria and fungi. However, haploids exist only in male and female gametes in mammals. But male and female gametes cannot be cultured *in vitro* for a long time, thus limiting their application in genetic screening. After successful establishment of mouse embryonic stem cells, scientists have been trying to establish haploid stem cells from mice or other species, but have not been successful. Until 2011, British scientists obtained stable haploid stem cells through continuous flow sorting, and these haploid cells were pluripotent and could differentiate into the various germ layers [32] both *in vitro* and *in vivo*. But haploid stem cells established in these studies were parthenogenetic haploid stem cells. Can mice haploids be established in a similar way? Chinese scientists, through micro-operation, removed the nucleus of an egg cell and injected sperm into it, and the resulting cells that were obtained were mouse androgenetic haploid-derived stem cells, and the most interesting feature of these cells is that these stem cells can replace the haploid sperm during oocyte fertilization and development of mouse [33, 34]. So, these androgenetic haploid stem cells not only have advantages in genetic screening but also due to their capability to replace the sperm can rapidly enable the formation of a transgenic animal. Chinese scientists have understood the haploid establishment and culture technology, and by the rapid application of this technology to other species have established cynomolgus monkey parthenogenetic haploid stem cells, and rat parthenogenetic and androgenetic haploid stem cells, all of which have a lot of advantages (Refs. [35, 36]) in the aspects of genetic screening, thus greatly expanding the application range of haploid stem cells and promoting rapid progress in this field. Haploid stem cells, as a new field, still have many unsolved problems.

## 5. Genetically Modified Animal Model and Gene Therapy

The development of new genetic tools is of great significance for the study of gene function and the field of regenerative medicine. In recent years, developments have included using the zinc finger gene editing enzyme, TALENs, and CRISPR/Cas technique. The CRISPR/Cas technique is used because of its advantages of high efficiency, simple design, and easy operation, and this has begun to replace other gene editing tools, and has now occupied the central position in gene editing operation

[37]. CRISPR/Cas itself is a defense system that protects bacteria and palaeophytes from the virus. The application of CRISPR/Cas to non-bacterial cells is based on two conditions. One is the Cas enzyme responsible for cutting the specific location of the genome. The other is the RNA (gRNA), which is responsible for identifying the specific location in the genome. In 2013, scientists applied this system to the genome of mammals and achieved successful results [38, 39]. Chinese scientists were the first to insert the prokaryotic mRNA into the fertilized egg of rats, and multiple gene knockout [40] in rats has also been successfully realized. Scientists also applied this technology for the treatment of cataract in mice. When injecting the CRISPR/Cas9 into the fertilized eggs, we restored the Crygc mutation of the mouse cataract-related gene, which allowed the mice to achieve normal vision again [41]. In 2014, Chinese scientists used CRISPR/Cas9 technology to obtain transgenic monkey and pig models, and this work was published in journals such as *Cell* [42, 43]. This is the first time that a genetic primate animal was created using CRISPR/Cas9 technology, and this highly economic method of creating a modified animal model caused a great sensation in the international research world. CRISPR/Cas9 can achieve efficient gene targeting in cells. Therefore, in the future, combining the repair of iPSCs with CRISPR/Cas9 will help promote cell differentiation and aid in transplantation, thus advancing the application of iPSCs in regenerative medicine.

In addition to these, DNA demethylation as well as the combination of adult stem cells and biomaterials are used to treat infertility and other conditions.


## 6. Future Prospects

After decades of development, stem cell technology has seen breakthroughs and innovations. Especially in recent years, the advent of induced pluripotent stem cell technology has laid the foundation for the basic research and application of stem cells. For a variety of genetic diseases (such as premature birth, sickle cell anemia, and schizophrenia) that occur in patients [44–46], iPSCs can well be used to simulate the disease *in vitro* after differentiation *in vitro*, thereby allowing scientists to observe the complete process of these diseases, which helps in further understanding of the pathogenesis of these diseases, studying the corresponding therapeutic drug development process, and also serving as a good reference. This work is now becoming increasingly important and catching more people's attention as there are many complex diseases of unknown origin that can be studied using disease-specific iPSCs to explore the etiology and treatment; therefore, the method of using disease-specific iPSCs to study diseases has continued to receive much attention. One of the important reasons why iPSCs are considered so important is that we hope to use this technology to cure human diseases one day. iPSCs can be applied as seed cells in the field of regenerative medicine.

Stem cell research has entered a critical phase of transforming from basic experimental research to clinical treatment. In 2011, the Food and Drug Administration (FDA) approved the Geron phase I clinical trial of Geron oligodendrocytes for the treatment of acute spinal cord injury.

In 2012, Health Canada approved the production and sale of the stem cell drug Osiris; current clinical studies have shown that hESCs differentiation to RPE cells is safe and effective for the treatment of AMD, and of 18 transplant patients 13 demonstrated visual improvement [47, 48]. In 2013, the Japanese Ministry of Health approved the clinical research for retinal regeneration using iPSCs. However, most of the cell lines currently registered at the National Institutes of Health (NIH) cannot be applied in the clinic [49] because they do not strictly meet the clinical requirements. So, the use of stem cells needs to undergo a rigorous inspection process; first of all, any stem cell research is required to obtain the approval of the ethics committee, and the source of the cell should be clearly specified; right from pre-clinical training, the whole process should be carried out at clinical-level conditions, and testing for endotoxin, mycoplasma, and human and animal sources of virus should be carried out as well as other safety testing procedures to ensure that the stem cells are of high purity, not contaminated, and not tumorigenic. Also, all the processing must be done under the review of the relevant state departments. At the same time, due to the need to repair the genetic variants associated with some genetic diseases, it is imperative to develop more secure and efficient gene editing methods. Quickly and efficiently developing research on the safety and effectiveness of iPSCs will have a positive impact on treatment in the field of regenerative medicine.

Pluripotent stem cell therapy is the only choice for regenerative medicine, as it is only this group of cells that can lead to the production of different, important functional cells through the process of trans-differentiation, and it is only these cells that are able to function properly *in vivo*. Therefore, the trans-differentiation of functional cells is also an important choice that has to be taken into consideration for medical regenerative treatment. At the same time, trans-differentiated cells obtained *in vitro* have better functions, and it is also more easy for them to establish contact with other cells [50–52]. But, trans-differentiation as a process is inefficient, the experimental system is not perfect, and in *in vivo* gene transfection orientation to a particular lineage is difficult. Therefore, if these problems can be solved, trans-differentiation can be used more efficiently, and this will surely be the answer to the prayers of patients who are anxious for cell transplantation.

However, compared with advanced international level of research, the basic research of stem cells in China is still very weak. Hence, there is fierce competition in the international stem cell field. Exploring mechanisms like reprogramming and mining the pluripotency of stem cells to maintain the network to improve stem cell differentiation capacity are trends that will continue for a long period of time, and these have currently become the research hotspots in the field of stem cells. If China can continue to increase investment in the field of stem cells, then it will achieve greater breakthroughs in the direction of basic research on stem cells in the future.

To sum up, stem cell research is a hot topic in the current international life science competition, and it is of great significance to the field of regenerative medicine in China that helps to enhance the competitiveness of the pharmaceutical industry. At present, the international stem cell research is in the early stages of development. China is facing unprecedented opportunities and has also achieved many important results. But facing the fierce competition in the international stem cell field, China is also looking at some unprecedented challenges. In order to make our clinical application of stem cells more run-of-the-mill and reasonable, our government departments have also developed a stem cell management system (stem cell clinical trial management measures (Trial), stem cell clinical trial base management measures (Trial), stem cell preparation quality control and clinical study of the guiding principles (Trial draft)). We believe that in the near future, with the introduction of specific policies, the clinical application of stem cells in China will launch a new era, contributing to the health of the majority of the people.

## References

1. Takahashi K, Yamanaka S. Induction of pluripotent stem cells from mouse embryonic and adult fibroblast cultures by defined factors. *Cell*, 2006, 126(4): 663–676.
2. Zhao XY, Li W, Lü Z, *et al.* iPS cells produce viable mice through tetraploid complementation. *Nature*, 2009, 461(7260): 86–90.

3. Kang L, Wang JL, Zhang Y, *et al.* iPS cells can support full-term development of tetraploid blastocyst-complemented embryos. *Cell Stem Cell*, 2009, 5(2): 135–138.

4. Liu L, Luo GZ, Yang W, *et al.* Activation of the imprinted *Dlk1-Dio3* region correlates with pluripotency levels of mouse stem cells. *J Biol Chem*, 2010, 285(25): 19483–19490.

5. Esteban MA, Wang T, Qin BM, *et al.* Vitamin C enhances the generation of mouse and human induced pluripotent stem cells. *Cell Stem Cell*, 2010, 6(1): 71–79.

6. Wang T, Chen KS, Zeng XM, *et al.* The histone demethylases Jhdm1a/1b enhance somatic cell reprogramming in a vitamin-C-dependent manner. *Cell Stem Cell*, 2011, 9(6): 575–587.

7. Chen JK, Guo L, Zhang L, *et al.* Vitamin C modulates TET1 function during somatic cell reprogramming. *Nat Genet*, 2013, 45(12): 1504–1509.

8. Stadtfeld M, Apostolou E, Ferrari F, *et al.* Ascorbic acid prevents loss of *Dlk1-Dio3* imprinting and facilitates generation of all-iPS cell mice from terminally differentiated B cells. *Nat Genet*, 2012, 44(4): 398–405, S1–S2.

9. Hou PP, Li YQ, Zhang X, *et al.* Pluripotent stem cells induced from mouse somatic cells by small-molecule compounds. *Science*, 2013, 341(6146): 651–654.

10. Shu J, Wu C, Wu YT, *et al.* Induction of pluripotency in mouse somatic cells with lineage specifiers. *Cell*, 2013, 153(5): 963–975.

11. Etzion S, Barbash IM, Feinberg MS, *et al.* Cellular cardiomyoplasty of cardiac fibroblasts by adenoviral delivery of MyoD *ex vivo*: An unlimited source of cells for myocardial repair. *Circulation*, 2002, 106(12 Suppl 1): I125–I130.

12. Sheng C, Zheng QY, Wu JY, *et al.* Direct reprogramming of Sertoli cells into multipotent neural stem cells by defined factors. *Cell Res*, 2012, 22(1): 208–218.

13. Huang PY, He ZY, Ji SY, *et al.* Induction of functional hepatocyte-like cells from mouse fibroblasts by defined factors. *Nature*, 2011, 475(7356): 386–389.

14. Yu B, He ZY, You P, *et al.* Reprogramming fibroblasts into bipotential hepatic stem cells by defined factors. *Cell Stem Cell*, 2013, 13(3): 328–340.

15. Huang PY, Zhang LD, Gao YM, *et al.* Direct reprogramming of human fibroblasts to functional and expandable hepatocytes. *Cell Stem Cell*, 2014, 14(3): 370–384.

16. Du YY, Wang JL, Jia J, *et al.* Human hepatocytes with drug metabolic function induced from fibroblasts by lineage reprogramming. *Cell Stem Cell*, 2014, 14(3): 394–403.

17. Cheng L, Hu WX, Qiu BL, *et al.* Generation of neural progenitor cells by chemical cocktails and hypoxia. *Cell Res*, 2014, 24(6): 665–679.

18. Wu L, Cai XX, Zhang S, *et al.* Regeneration of articular cartilage by adipose tissue derived mesenchymal stem cells: Perspectives from stem cell biology and molecular medicine. *J Cell Physiol*, 2013, 228(5): 938–944.

19. Keating A. Mesenchymal stromal cells: New directions. *Cell Stem Cell*, 2012, 10(6): 709–716.

20. Miller RH, Bai L, Lennon DP, *et al.* The potential of mesenchymal stem cells for neural repair. *Discov Med*, 2010, 9(46): 236–242.

21. Xu GW, Zhang YY, Zhang LY, *et al.* The role of IL-6 in inhibition of lymphocyte apoptosis by mesenchymal stem cells. *Biochem Biophys Res Commun*, 2007, 361(3): 745–750.

22. Xu GW, Zhang LY, Ren GW, *et al.* Immunosuppressive properties of cloned bone marrow mesenchymal stem cells. *Cell Res*, 2007, 17(3): 240–248.

23. Park D, Yang G, Bae DK, *et al.* Human adipose tissue-derived mesenchymal stem cells improve cognitive function and physical activity in ageing mice. *J Neurosci Res*, 2013, 91(5): 660–670.

24. Sun LY, Akiyama K, Zhang HY, *et al.* Mesenchymal stem cell transplantation reverses multiorgan dysfunction in systemic lupus erythematosus mice and humans. *Stem Cells*, 2009, 27(6): 1421–1432.

25. Shen Q, Chen B, Xiao Z, *et al.* Paracrine factors from mesenchymal stem cells attenuate epithelial injury and lung fibrosis. *Mol Med Rep*, 2015, 11(4): 2831–2837.

26. Shi Q, Gao W, Han XL, *et al.* Collagen scaffolds modified with collagen-binding bFGF promotes the neural regeneration in a rat hemisected spinal cord injury model. *Sci China Life Sci*, 2014, 57(2): 232–240.

27. Lin NC, Li XA, Song TR, *et al.* The effect of collagen-binding vascular endothelial growth factor on the remodeling of scarred rat uterus following full-thickness injury. *Biomaterials*, 2012, 33(6): 1801–1807.

28. Han SF, Zhao YN, Xiao ZF, *et al.* The three-dimensional collagen scaffold improves the stemness of rat bone marrow mesenchymal stem cells. *J Genet Genomics*, 2012, 39(12): 633–641.

29. Wei JS, Han J, Zhao YN, *et al.* The importance of three-dimensional scaffold structure on stemness maintenance of mouse embryonic stem cells. *Biomaterials*, 2014, 35(27): 7724–7733.

30. Guan J, Zhu ZH, Zhao RC, *et al.* Transplantation of human mesenchymal stem cells loaded on collagen scaffolds for the treatment of traumatic brain injury in rats. *Biomaterials*, 2013, 34(24): 5937–5946.

31. Higuchi A, Ling QD, Chang Y, *et al.* Physical cues of biomaterials guide stem cell differentiation fate. *Chem Rev*, 2013, 113(5): 3297–3328.

32. Leeb M, Wutz A. Derivation of haploid embryonic stem cells from mouse embryos. *Nature*, 2011, 479(7371): 131–134.

33. Yang H, Shi LY, Wang BA, *et al.* Generation of genetically modified mice by oocyte injection of androgenetic haploid embryonic stem cells. *Cell*, 2012, 149(3): 605–617.

34. Li W, Shuai L, Wan HF, *et al.* Androgenetic haploid embryonic stem cells produce live transgenic mice. *Nature*, 2012, 490(7420): 407–411.

35. Yang H, Liu Z, Ma Y, *et al.* Generation of haploid embryonic stem cells from Macaca fascicularis monkey parthenotes. *Cell Res*, 2013, 23(10): 1187–1200.

36. Li W, Li X, Li TD, *et al.* Genetic modification and screening in rat using haploid embryonic stem cells. *Cell Stem Cell*, 2014, 14(3): 404–414.

37. Doudna JA, Charpentier E. The new frontier of genome engineering with CRISPR-Cas9. *Science*, 2014, 346(6213): 1258096.

38. Cong L, Ran FA, Cox D, *et al.* Multiplex genome engineering using CRISPR/Cas systems. *Science*, 2013, 339(6121): 819–823.

39. Mali P, Yang LH, Esvelt KM, *et al.* RNA-guided human genome engineering *via* Cas9. *Science*, 2013, 339(6121): 823–826.

40. Li W, Teng F, Li TD, *et al.* Simultaneous generation and germline transmission of multiple gene mutations in rat using CRISPR-Cas systems. *Nat Biotechnol*, 2013, 31(8): 684–686.

41. Wu YX, Liang D, Wang YH, *et al.* Correction of a genetic disease in mouse *via* use of CRISPR-Cas9. *Cell Stem Cell*, 2013, 13(6): 659–662.

42. Niu YY, Shen B, Cui YQ, *et al.* Generation of gene-modified cynomolgus monkey *via* Cas9/RNA-mediated gene targeting in one-cell embryos. *Cell*, 2014, 156(4): 836–843.

43. Hai T, Teng F, Guo RF, *et al.* One-step generation of knockout pigs by zygote injection of CRISPR/Cas system. *Cell Res*, 2014, 24(3): 372–375.

44. Liu GH, Barkho BZ, Ruiz S, *et al.* Recapitulation of premature ageing with iPSCs from Hutchinson-Gilford progeria syndrome. *Nature*, 2011, 472(7342): 221–225.

45. Hanna J, Wernig M, Markoulaki S, *et al.* Treatment of sickle cell anemia mouse model with iPS cells generated from autologous skin. *Science*, 2007, 318(5858): 1920–1923.

46. Brennand KJ, Simone A, Jou J, *et al.* Modelling schizophrenia using human induced pluripotent stem cells. *Nature*, 2011, 473(7346): 221–225.

47. Schwartz SD, Hubschman JP, Heiwell G, *et al.* Embryonic stem cell trials for macular degeneration: A preliminary report. *Lancet*, 2012, 379(9817): 713–720.

48. Schwartz SD, Regillo CD, Lam BL, *et al.* Human embryonic stem cell-derived retinal pigment epithelium in patients with age-related macular degeneration and Stargardt's macular dystrophy: Follow-up of two open-label phase 1/2 studies. *Lancet*, 2015, 385(9967): 509–516.

49. Jonlin EC. Differing standards for the NIH Stem Cell Registry and FDA approval render most federally funded hESC lines unsuitable for clinical use. *Cell Stem Cell*, 2014, 14(2): 139–140.

50. Rouaux C, Arlotta P. Direct lineage reprogramming of post-mitotic callosal neurons into corticofugal neurons *in vivo*. *Nat Cell Biol*, 2013, 15(2): 214–221.

51. Qian L, Huang Y, Spencer CI, *et al. In vivo* reprogramming of murine cardiac fibroblasts into induced cardiomyocytes. *Nature*, 2012, 485(7400): 593–598.

52. Zhou Q, Brown J, Kanarek A, *et al. In vivo* reprogramming of adult pancreatic exocrine cells to beta-cells. *Nature*, 2008, 455(7213): 627–632.

# Chapter 3

# The Structural and Functional Characteristics of the Intervertebral Disc

The intervertebral disc is a part of the human spine and has the most important part to play in the system, functioning as a shock absorber, as it slows down the shock and distributes the force uniformly, provides the longitudinal axis of the spine, and helps maintain the stability of spine in a range of activities, apart from other important biomechanical functions. Stress plays an important role in the formation of the nucleus pulposus and annulus cells during the development of the intervertebral disc. The structural characteristics of the disc depend on the nature of its components, and the intervertebral disc tissue is influenced by the human torso and upper limb weight, as it bears a heavier strain than the other tissues. Long-term load bearing process has an important influence on the morphology, phenotype, and proliferation, as well as on the synthesis of collagen and protein polysaccharides and other matrix components that play a role in intervertebral disc cells. The intervertebral disc is the body's largest nonblood transport organization system. The support of blood vessels is closed within a few months after birth. Nutrition mainly depends on the permeability of the cartilage endplate, which is easily permeable [1]. In addition, the cell division and proliferation capacity of nucleus pulposus and annulus cells are very poor, and the intervertebral disc tissue regeneration and repair capacity is also limited. All of these characteristics reiterate the necessity of intervertebral disc tissue transplantation.

## 1. Anatomical Structure of Intervertebral Disc

The intervertebral disc is composed of three parts: cartilage, annulus, and nucleus (Fig. 1).

(a) Cartilage endplate: The cartilage endplate is composed of fibrous cartilage, and the upper and lower part of the vertebral body, with an average thickness of 1 mm. There are many micro pores in the cartilage endplate, which is the pathway through which water and metabolites of the nucleus pulposus pass. From birth to 8 months of age, the capillaries begin to close, and during the age of 20–30 years, atresia starts. There is no nerve tissue in the cartilage endplate, and so when the cartilage endplate is injured, pain cannot be felt,

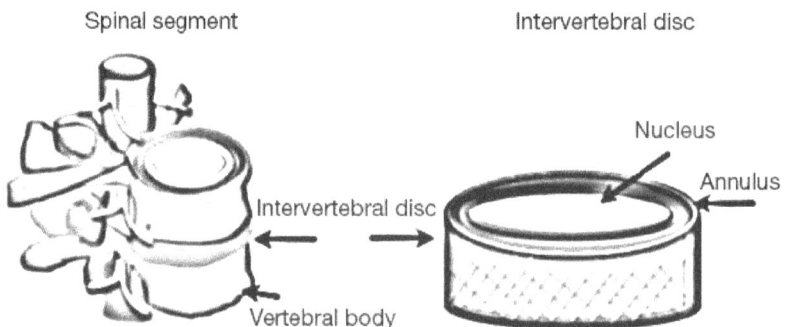

Fig. 1.   The intervertebral disc is composed of three parts: cartilage, annulus, and nucleus.

and it thus cannot repair itself [2, 3]. Cartilage in the form of articular cartilage can withstand pressure, thereby enabling the vertebra to be protected from pressure overload, and this helps in maintaining the integrity of vertebral cartilage end-plate, and the vertebral body will be protected from the pressure absorption phenomenon.

(b) The annulus fibrosus (fiber ring) is divided into three layers: outer, middle, and inner. The outer layer is composed of a collagen fiber band, and the inner layer is composed of a fibrous cartilage zone. Bonding between each layer helps create a solid structure. The front and side sections of the fiber ring are thick, almost two times that of the rear side. The innermost fibers are in the nucleus and are connected to the nucleus, so there is no clear demarcation between them. The whole fiber ring is almost a concentric circle, with the outer peripheral fiber more vertical, and the central angle of inclination greater. The fiber ring is very strong and is closely attached to the cartilage endplate, to help maintain the stability of the spine.

(c) The nucleus pulposus is relatively large and soft and is located at the center of the intervertebral disc, and it does not come into contact with the vertebral body. During the processes of growth and development, the position of nucleus pulposus changes. The rear side of the vertebral body progresses faster than the front, so the nucleus is located in the posterior side of the intervertebral disc. The nucleus pulposus accounts for 50–60% of the intervertebral disc when taken as a cross-section. In early childhood, the inner annulus of the intervertebral disc wraps around the notochord cells; once the notochord cells disappear, after the age of 10, only a soft tremellose nucleus pulposus is observed. At the age of 12, this is almost completely replaced by a loose fibrous cartilage and a large number of collagen components. As aging progresses, the collagen material is gradually replaced by fibrous cartilage [4].

In children, the structure of nucleus pulposus and the fiber ring are obvious, but in old age, the water content of nucleus pulposus is decreased and the boundary between the fibrous cartilage and nucleus pulposus is not obvious. The differentiation in the structure of the nucleus pulposus collagen fiber network is relatively dense, which is not good. Each layer of collagen fibers is coated with a mucopolysaccharide protein complex and chondroitin sulfate, and the nucleus has the ability to combine with water. Depending on a person's age, the water content of nucleus pulposus can account for 75–90% of the total nucleus [5]. The various components of the nucleus pulposus are combined together to form a 3D network of rubber-like structures. With the changes in the height of normal people and changes in the water content in the nucleus pulposus, the fibrous cartilage from the fibrous ring and the cartilage plate are gradually replaced by the fibrous collagen in the nucleus pulposus, and the morphology of the nucleus pulposus changes. The nucleus pulposus is malleable, and pressure causes it to become flat, so that the pressure can be transmitted in each direction. The nuclei of adjacent vertebral bodies also play a supporting role, like a ball, moving forward or backward with the flexion and extension of the spine.

## 2. Lumbar Intervertebral Disc

### 2.1. *Lumbar intervertebral disc, intervertebral foramen, and nerve root*

The spinal dorsal root nerve fibers and ventral root nerve fibers in the dorsal root ganglia combine together, become a hybrid neural stem, and exit the spinal canal through the intervertebral foramen (Fig. 2). Most of the lumbar nerve dorsal root ganglion is in the intervertebral foramen, but the dorsal root ganglion of the sacral nerve is located in the sacral canal. The lumbar nerve is divided into dorsal and ventral branches. The

# Intervertebral Disc

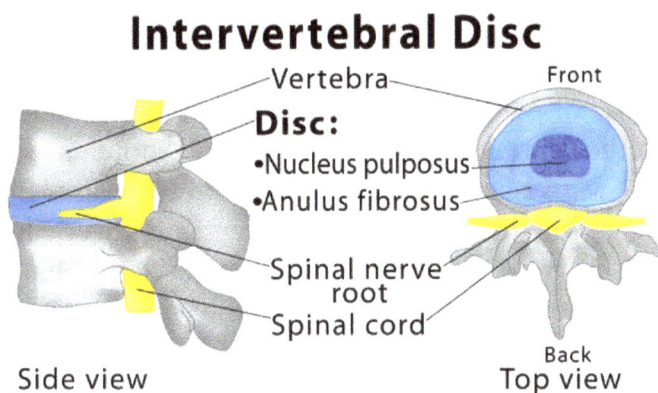

Fig. 2. Lumbar intervertebral disc, intervertebral foramen, and nerve root. Spinal dorsal root nerve fibers and ventral root nerve fibers in the dorsal root ganglia combine together, become a hybrid neural stem, and come out of the spinal canal through the intervertebral foramen. Image courtesy of ehealthstar.com [33]

nerve root in the cervical intervertebral foramen is most easily compressed. The upper and lower intervertebral foramen have large diameters. So, when there is protrusion of intervertebral disc and facet joint synovial swelling, bone hyperplasia can lead to intervertebral foramen stenosis, making the opening smaller than the diameter of the nerve root, leading to symptoms of nerve root compression [6–8]. Under normal circumstances, the following can be observed: the waist's lumbar disc (L3 and L4) herniation, compression of the 4th lumbar nerve root, lumbar disc herniation (L4 and L5), oppression of the 5th lumbar nerve root, disc herniation of L5 and S1, and compression of the 1st sacral nerve root.

## 2.2. *The relationship between lumbar intervertebral disc and adjacent structures*

The purpose of understanding the adjacent structures is to avoid the damage to these structures during interventional therapies and surgical treatment, so as to prevent serious complications.

The anterior part of the vertebral body and the intervertebral disc form the central part of the posterior abdominal wall. The abdominal aorta is in contact with intervertebral discs L1–3, the abdominal aorta with the 4th lumbar centrum lower edge to the bifurcation of common iliac artery, and the left common iliac artery on the left side with the 4th lumbar intervertebral disc. The anterior longitudinal ligament is attached and covered in the front of the vertebral body and intervertebral disc. The posterior structure of the lumbar intervertebral disc and the front wall of the vertebral canal are formed together with the vertebral body. The longitudinal ligament connects the intervertebral disc and the posterior central part, and the posterior longitudinal ligament on both sides is thus strengthened, so intervertebral disc herniation occurs on the side. Spinal artery and vein and nerve fibers are seen in the rear of the intervertebral disc.

## 3. Intervertebral Disc Nutrition

The adult intervertebral disc is the largest nonvascular tissue in the body, and its own nutritional and metabolic products are processed through the intervertebral disc outside the blood vessels [9]. In the outer layer is the nutrient supply around the vertebral body, which is delivered by the vertebral artery of small blood vessels. The cartilage disc nutrition supply relies on direct contact with vertebral bone marrow, whereas nutrition supply to the nucleus occurs due to the permeability of cartilage endplate (Fig. 3).

### 3.1. *Factors affecting the nutrition of intervertebral disc*

(a) Any potential risk factors to the nutritional supply of the intervertebral disc can be affected by the interference of any

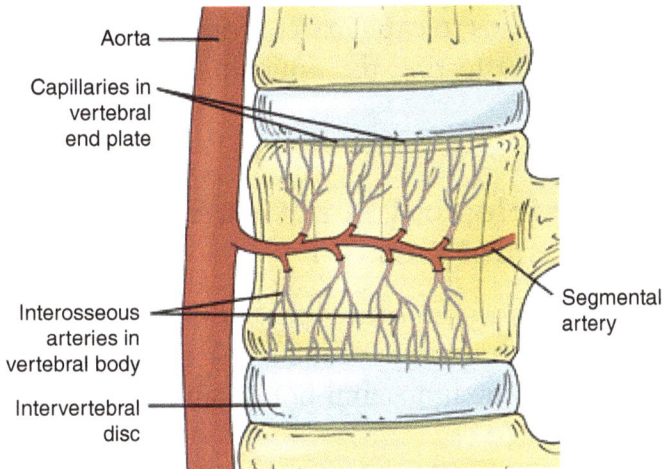

Fig. 3. The intervertebral disc is the largest nonvascular tissue in the body, and it has its own nutritional and metabolic products that are processed through the intervertebral disc outside the blood vessels. In the outer layer is the nutrient supply around the vertebral body which is due to the vertebral artery of small blood vessels; cartilage disc nutrition supply relies on direct contact with vertebral cancellous bone marrow, whereas the nutrition supply to the nucleus is through permeability of cartilage endplate. Reprinted with permission from clinicalgate.com [34]

part of the capillary network surrounding the intervertebral disc.

(b) Exercise can improve the nourishment of the intervertebral disc, but can also damage it. It is difficult to predict the effect of exercise, but it is generally believed that a particular amount of exercise may be useful.

(c) The fusion of intervertebral disc segments adjacent to the fused brake disc decreases metabolic activity of the intervertebral disc, leading to some cell death.

(d) Shock to the spine and the intervertebral disc system and any special movement bearing excessive pressure will have a negative impact on the structure, cells, and molecules of the intervertebral disc, which can lead to decreased water content and sulfate uptake apart from a decrease in disc height.

(e) Smoking leads to capillary contraction and a decrease in blood flow, thus affecting the nourishment of the intervertebral disc.

## 4. The Physiological Function of the Intervertebral Disc

The spine is the most important part of the entire exercise system, which bears the weight of the trunk of the body, serves as the backbone for the limbs and bones, and protects the spinal cord and spinal nerves.

The main functions of the intervertebral disc are as follows:

- To maintain the height of the spine: The vertebral body develops and intervertebral disc grows, in order to increase the length of the spine.
- To form a connection between the adjacent vertebra, both above and below it, to provide a certain degree of activity.
- The same force-bearing surface of the vertebral bodies, and the even interbody elements function through the nucleus' semi-liquid ingredients. The entire disc is subjected to the same stress.
- As a kind of buffer: Due to the elastic structure, especially of the nucleus pulposus, the nucleus is flexible, and the pressure causes it to be flat, so that the force applied to the nucleus can be compressed. The average force can be transferred to the fibrous ring and the cartilage in each direction, which are the main structures of spinal shock absorption and play the role of an elastic cushion. This is especially the case when a person falls onto their shoulder, back, or waist suddenly, leading to transmission of the load buffer so as to protect the spinal cord and brain nerves.
- To maintain a certain distance from the lateral facet and provide height.

- To maintain the size of the intervertebral foramen, which is usually 3 to 10 times the diameter of the nerve root.
- To maintain the spinal curvature and keep intervertebral disc thickness of different parts of lumbar intervertebral disc the same, that is, the thickness of the front and rear, which could lead to lumbar physiological lordosis.

The cartilage endplate functions to cover the vertebral body, and the vertebral compression bone atrophy occurs due to pressure; also, between the cartilage endplate and intervertebral disc, there is a soft bone marrow where nutrient exchange occurs. The function of nucleus pulposus is for absorption, oscillation, conduction, and maintenance of equilibrium pressure. The fiber ring connects the upper and lower vertebral body, maintains the stability of spine motion, prevents excessive spinal activity, keeps the liquid composition of the nucleus to maintain the position and shape of nucleus pulposus, and also helps shock absorption.

## 5. Biochemical Changes in Intervertebral Disc Degeneration

### 5.1. *Changes in collagen content*

Intervertebral disc contains type I and type II collagen, of which 60% is type II collagen and 40% type I collagen [10, 11]. The physical properties of collagen fibers change with age, such as decrease in shrinkage and tensile strength. There was also a gradual emergence of fibrosis, which caused the nucleus pulposus and the fiber ring to fuse and the ability of nucleus transfer and equilibrium pressure to decrease.

### 5.2. *The change of protein polysaccharide levels*

The disc contains proteoglycan core protein, and it is attached to chondroitin sulfate and keratan sulfate glycosaminoglycan through hyaluronic acid polymerization, therefore leading to protein stability. Along with increasing age and degeneration, a decrease is seen in the total protein content of polysaccharide, keratan sulfate, and chondroitin sulfate. These changes are not only related to the lack of nutrient sources in the intervertebral disc, but also to the biological characteristics of the chondrocytes.

## 5.3. *The change of water content*

The water content of the normal intervertebral disc ranges between 85% and 78%, and the content of water can decrease to 70%. Lumbar disc herniation is one of the most common clinical disorders of the waist. Trauma and strain cause the supporting intervertebral disc annulus to become weak, or even rupture. The fibrous ring is weak or ruptures more in the rear or side of the intervertebral disc [12–15]. The lumbar intervertebral disc to the rear of the spinal canal is prominent, and its rupture or prolapse can lead the adjacent tissues, such as spinal nerve root, to suffer from oppression, resulting in lower back pain, lower extremity numbness and pain, and other clinical symptoms. Physiological degeneration is related to age and its associated biological changes, and this results in increased force, which leads to brittleness. But there is no definite boundary between the physiological process and the pathological process. The degeneration begins after 20 years of age, annulus fibrosus and nucleus pulposus are 70–80% aqueous in nature, and they degenerate gradually after the loss of moisture and lack of nutrition. After the collapse, the intervertebral disc tissue is 1/4 the original volume [16–19].

When there is a protrusion of the tissue surface, there are vascular wounds around the invasion, resulting in inflammation, and ultimately leading to prominent tissue fibrosis and calcification. Fibrosis and calcification can extend to the fiber ring. X-ray diffraction electron microscopy revealed that the main component of this pathological calcium deposit was hydroxyapatite [20].

The calcification of the intervertebral disc in children showed different pathological processes, which can be divided into three types: regression type, dormancy type, and static type [21]. The regression type of intervertebral disc calcification is related to the calcification deposition in the nucleus pulposus, but the more common one is the invasion of cervical intervertebral disc; lumbar spine area involvement is rare. Dormancy type can be found on spine X-ray. There may be severe symptoms, but the symptoms of calcification deposition can disappear. In the rest of the intervertebral disc, calcification can occasionally be found as interstitial calcification, and signs and symptoms are not related.

The Schmorl nodule is the nucleus pulposus' upward or downward protrusion through the cartilage plate vertically into the vertebral body. Schmorl nodules were found at autopsy in 38% of people, occurring in the thoracic and lumbar spine in 39.9% males and 34.3% females. Schmorl nodules were seen to be prevalent, at almost twice the rate, in women aged 60 years or older compared to men. Andrae and co-workers, in 368 cases of autopsy, found that 15% of the people had Schmorl nodules in the vertebral body and spinal canal [22]. An anatomic study recently found that the majority of Schmorl nodules occur in young people and a relatively small proportion of adults, suggesting that nucleus pulposus prominent material is in a semi-liquid form and that, at least in some cases, there is already prominent degeneration present in the region [23].

The majority of Schmorl nodules of diameter below 5 mm have a mushroom-like shape and are seen in the center of the posterior vertebral body. The movement of the nucleus into the vertebral body can lead to the following changes: vertebral trabecular bone fracture and local necrosis, formation of a cavity to accommodate the prominent nucleus, inflammation around the nucleus, and absorption of the necrotic tissue; the nucleus pulposus can gradually increase or do so rapidly, and this is based on the pressure balance around the outburst from the spinal cord; nuclear dehydration causes the metaplastic cartilage or bone cells surrounding the nucleus and the trabecular bone to increase in density, limiting the further expansion of the blood vessels around the nucleus pulposus. The cartilage fracture into the intervertebral disc leads to fibrosis, calcification, and ossification of the nucleus. The nodule can be clearly seen on an X-ray film at this time.

## 6. The Pathological Classification and Clinical Classification of Lumbar Disc Herniation

The pathological processes seen in the intervertebral disc are mainly in the form of biochemical changes in the nucleus. Once the nucleus disappears completely, a serious break in the functioning of the unit is observed. The pathological process of lumbar disc herniation can be divided into three stages, and the pathological changes of the intervertebral disc can lead to alterations in the tissue, which results in obvious symptoms and signs. Three pathological stages proposed by Amstrong are as follows: the progressive degeneration of nucleus pulposus, the reduction of water, and the formation of necrotic mass. In detail, fiber degeneration and posterior ring fracture are seen; the nucleus pulposus degenerates due to the presence of the broken fibers in the ring; eventually, the fiber ring is replaced by the fibrous connective tissue of the intervertebral disc. The intervertebral disc space also narrows. However, most of these stages are still being discussed. As early as the 1940s, Scitt and Young divided disc herniation into two types: fixed and active lumbar disc herniation. The Schmorl nodules can be divided into recurrent herniation, fixed protrusion, incarcerated and free of protrusion [24]. Mcnab classified the protrusion of the intervertebral disc into four stages. The peripheral annular bulge, wherein the posterior ring fiber is more prominent, which does not cause serious nerve root compression. The annular bulge, where there is protrusion of the intervertebral disc nucleus pulposus, which shifts a few layers to within the annulus. The intervertebral disc prolapse, which is marked by the displacement of the nucleus pulposus through the fiber ring and is located in the posterior longitudinal ligament. The intervertebral disc-free material case, in which the prolapse of lumbar intervertebral disc nucleus pulposus occurs free from spinal or epidural material, intervertebral foramen, etc.

Song Xianwen, after performing operations and observing the prominences in the intervertebral discs, proposed the following classification: complete type, with outer annular integrity and spherical nucleus; the rupture of subperiosteal type, where fiber ring can completely disappear, protrusion is oblong, or can be ragged in the up or down direction to the adjacent vertebral body behind the spinal canal; implosive type, where fiber ring has broken out of the protrusion and it can be located in the posterior longitudinal ligament or free to the vertebral canal.

In the outstanding type of degeneration, the inner and outer fibrous rings rupture completely so that the nucleus pulposus escapes through the slit from the top and the outer bulge. Schmorl and Junghanns considered that there was prominent substance in addition to nucleus pulposus. These are the fibrous ring and cartilage endplate. During the operation, majority of the disc herniation is found to be due to the limitation of the hemispherical bulge of the fiber ring, or fiber ring has been broken. The fibrous ring and nucleus pulposus tissue are prominent in the posterior longitudinal ligament. In young people, it is more than a part of the cartilage endplate, as it is involved with some of the vertebral body backward shift. In epidural space or intervertebral foramen, and even into the epidural cavity, cartilage tumor-like material is seen. Lindblom defined statistical posterior disc protrusion seen in different people as follows: fiber ring relaxation, surface fiber degeneration after bulging, ring still complete — 43.2%; the fiber ring breaking, the covering of the surface of nucleus pulposus becomes prominent with complete posterior longitudinal ligament — 25.3%; presence of nucleus pulposus in the spinal canal — 26.6%; intervertebral disc atrophy, or scarring, and nerve root adhesion — 2.8%; the nucleus pulposus degeneration tension being lower than the normal — 2.1% [25–27]. Based on the nucleus' position in the lumbar disc, Wu Zuyao divided herniation into the following forms: intervertebral disc protrusion; a few fractures formed in the fiber ring; small degree and large degree outburst; outburst of the nucleus pulposus; and overall degeneration. But not every case of lumbar disc herniation follows this process. Some of them express in one or several forms.

The size of the protrusion and prominent direction: the prominence can range from having a small diameter of only 5–6 mm, similar in size to soybean, to having a large diameter of up to 1 cm — finger-like prominence into the spinal canal. It has been reported that the maximum weight of the protrusion can be up to 5 g, but we have found a maximum weight of about 15 g in our clinical work. But the relationship between the size of the protrusion and the symptoms is not completely consistent, and the symptoms are related to the size and shape of the spinal canal. Especially in the three-leaf shaped spinal canal, where the lateral recess is narrow, the small prominence can produce obvious symptoms. The proportion of protrusion of the intervertebral disc and spinal canal varies according to the size of the local diameter, and it is divided into three degrees: mild protrusion — highly local, does not exceed the anteroposterior diameter of the spinal canal by 1/3; moderate — does not exceed the local anteroposterior diameter of the spinal canal by 1/2; severe — exceeds the anteroposterior diameter of the spinal canal by 1/2 [28].

The relationship between the clinical classification and the pathological type, as provided by Spengler, is divided into three types: convex type (protruded), prominent type (extruded), and free type (sequestered). Song Xianwen's classification is similar, and is more closely linked to clinical surgery classification, and this can help to show the degree of rupture, thereby enabling clinical applications to be carried out with ease. Depalma and Rothman also classified lumbar disc herniation into three types according to the position of the prominence, namely, lateral side prominence, prominence in the nerve hole, and central prominence. The pathological changes seen in the different types were also put forward. The posterior lateral prominent fiber ring is behind the weakest part in the midline of the intervertebral disc. Due to the degeneration of the intervertebral disc, collagen type II levels are increased, the fiber ring itself becomes weak, and the posterior longitudinal ligament lacks the strength to reinforce the central fiber. Therefore, the most common site for lumbar disc herniation depends on the size and the pressure level of the intervertebral disc nucleus pulposus tissue in the posterior longitudinal ligament. At this time, the outstanding prominence is a hard and smooth uplifted structure. And the posterior longitudinal ligament is separated from the vertebral body. When the size increases, thus leading to further separation of vertebral ligament, nucleus pulposus tissue can move in any direction, but usually does so in the middle or lateral direction, to the nerve root, or parallel to the direction of the protuberance into the intervertebral foramen. The nucleus can be completely dissociated or can still be connected with the fibrous tissue in the nucleus pulposus. This type is the most common one. Any point where the dural sac through the intervertebral foramen comes into contact with the nerve root can lead to a prominence in the nucleus pulposus. In most cases, the material can be present directly in the nucleus or nerve root, or in the inner and outer sides of the nerve root, causing stretch and tension. This prominence can not only cause nerve root tension but also lead to compression of the nerve root to the osseous lamina or excessive wrinkle on the yellow ligament.

When spinal canal stenosis is observed (developmental or acquired), the chance of nerve root compression increases significantly. Some authors believe that excessive wrinkles in the yellow ligament usually do not cause oppression. But in Depalma and Rothman's opinion, this is a more common factor for compression [29, 30]. In hyperextension, the yellow ligament creases, decreases in volume, and can block the nerve root compression. As the lumbar sacral angle increases, especially close to the level of the lumbar spine, compression of the dural sac, and the formation of cyclic compression, is observed. At this time, the nerve roots are the main ones being compressed. The chief complaint of sensorimotor disorder is most often not pain; the intervertebral disc becomes prominent, and it protrudes out back through the posterior annulus and posterior longitudinal ligament into the spinal canal, or into the intervertebral foramen, ruptures gradually after which the posterior longitudinal ligament also becomes prominent and protrudes into the intervertebral foramen. In the intervertebral foramen, the protrusion can be compressed and the nerve root causes its stress. The straight leg raising test or supine abdominal test can produce severe radiative pain in the lower limbs. Stephens, observing 20 cases of lumbar spine specimens, found that in lumbar intervertebral disc abnormalities, foraminal deformation can lead to oppression of the nerve root and cause symptoms and signs. Thus, the central protrusion and central disc are really prominent, the nucleus material by the fiber ring is prominent, and longitudinal ligament is also visible. The posterior longitudinal ligament in the central part of the fiber thickens, strengthening the fiber ring, so the complete rupture of the outer layer of the fiber ring is rarely seen. The posterior longitudinal ligament causes extreme spinal flexion, which can cause rupture into the spinal canal through the nucleus pulposus material. In the central type of lumbar disc herniation, nerve root compression and local formation of fibrous connective tissue surrounding the nerve root are observed, but nerve root abnormalities, edema, and the formation of a strip of medullary sheath are also observed. The nucleus in the posterior longitudinal ligament can also lead to nerve root compression, resulting in symptoms.

## 7. The Mechanism of Leg Pain in Lumbar Disc Herniation

There are three main theories:

(1) Since the mechanical compression theory proposed by Mixter and Barr for the first time in 1934 for lumbar disc herniation surgery, many scholars have believed that mechanical compression of the nerve root is the main cause of low back pain and sciatica. Also, the affected nerve is pretty severely affected, and a simple nerve compression is rarely seen. In these cases, tension can often develop, and if this tension is not promptly removed, nerve inflammation and edema result, causing an increase in nerve tension, which leads to gradual increase in nerve dysfunction.

The spinal nerve has a rich outer membrane, wrapped around the nerve bundle. This outer membrane is composed of elastic collagen structure and adipose tissue, so it has an elastic cushioning effect, and thus the nerve is not easily damaged by mechanical forces. In the inner layer of the outer layer of the nerve, there is a nerve bundle layer, and this film has a chemical barrier function, which can protect from foreign chemical stimulation, thus safeguarding the nerve from chemical damage. But the nerve root of the nerve tissue is very underdeveloped, and so it possesses no elastic buffer and chemical barrier function. This makes it susceptible to mechanical and chemical damage. Therefore, nerve root injury is extremely common in lumbar disc herniation.

(2) The theory of chemical nerve root inflammation plays an important role as a possible causative factor of pain, but it cannot completely be taken as the source of pain. There is no pain during normal nerve compression; pain only occurs when inflammatory nerve compression is caused. The nerve roots in the vicinity of the herniated intervertebral disc often show signs of congestion, edema, and inflammatory changes. This kind of inflamed nerve is very sensitive to pain, and the normal functioning of the nerve can cause severe pain. With regard to the cause of nerve root inflammation, Mashall posits that the mechanism is mainly the degeneration of the intervertebral disc annulus, weak burst, liquid pulp karyolymph from crevasse overflow, and diffusion between the disc and nerve root channels. In the pulp, karyolymph glycoprotein and beta protein act as strong chemical irritants to the nerve root, while a large amount of "H" material (histamine release) is also present; thus, the nerve root and nerve bundle membrane serve as a chemical barrier, resulting in chemical radiculoneuritis. Murphy also pointed out that during inflammation, a variety of chemical factors can lead to increased vascular permeability to proteins, thereby leading to large amounts of histamine release. In the epineurium, perineurium, and endometrium, a large number of histamine-containing mast cells are seen, and in the nerve root, which is exudation prone, and sinus vertebral nerve, a large amount of inflammatory albumin is seen. This change leads to the increase of the internal pressure of the nerve, causing local ischemia and electrolyte disturbance, thus stimulating the nerve root and the sinus nerve, causing pain in this area. At the same

time, the local change in conduction may also damage the normal nerve and the formation of artificial synapse, leading to changes in the function of the synaptic active spinal nociceptive afferent fibers causing a short-circuit, which results in acute lumbar disc syndrome. Saal believed that this abnormal structure was not the only explanation for the mechanical barrier of nerve components leading to radicular pain caused by lumbar disc herniation, and therefore studied the inflammatory substance phospholipase A2 in the intervertebral disc specimens. He found that the activity was 2–10 times higher than that seen in any other area, and because this enzyme releases arachidonic acid, it limits the speed with which prostaglandin and interleukin three are generated [31]. Thus, the biological and chemical evidence of inflammation in the lumbar intervertebral disc protrusion was provided.

(3) In the past few years, Gertzbein and others, through a large number of animal experiments and clinical research, put forward the idea of the intervertebral disc. The intervertebral disc nucleus pulposus tissue is the largest of its kind that has no blood vessels and is also not in contact with the surrounding circulation, and so its nutrition is mainly derived from the diffusion through the cartilage disc. The body of the nucleus pulposus is protected from immune mechanism. With intervertebral disc injury or lesion, after the nucleus breaks through the longitudinal ligament fiber ring or during the repair process of neovascularization of the nucleus pulposus, it is in close contact with the immune mechanism. It is during this time that the glycoprotein and beta protein in the nucleus pulposus will become the antigen, thus leading to the body's response to such a continuous antigen stimulation, resulting in an immune response. Due to this immune response, a segment of the herniated disc can also cause degeneration of other segments of the intervertebral disc, leading to further pain. Gertzbein's classification and other methods such as

lymphocyte transformation test or leucocyte migration inhibition test can be used to measure the existence of cellular
immune response after intervertebral disc protrusion. Qingdao
Medical College has carried out related research similar to
this and has confirmed that the body does have its own
immune phenomenon [32]. The lumbar disc surgery patients
were studied by the immunology team, and it was found that
lumbar disc herniation leads to abnormal humoral immune
response, of which IgG and IgM might be acting as the antibody in the intervertebral disc tissue, thus being associated
with intervertebral disc degeneration. Also, IgA, IgD, and IgE
seem to play no roles in intervertebral disc degeneration. The
complement system may be involved in the autoimmune reaction against intervertebral disc tissue, in patients with chronic
low back pain, in the exclusion of other autoimmune diseases
and liver, kidney, and hematopoietic system diseases. One can
then detect IgG and IgM to try to assist in the diagnosis of the
immune disorder, especially in patients with ruptured or free
lumbar disc herniation. It has thus been found that there are
abnormal cellular immune responses seen in the lumbar disc
herniation.

# References

1. Buckwalter JA. Aging and degeneration of the human intervertebral
   disc. *Spine*, 1995, 2011: 1307–1314.
2. Fujiwara A, Lim TH, An HS, *et al*. The effect of disc degeneration and
   facet joint osteoarthritis on the segmental flexibility of the lumbar
   spine. *Spine*, 2000, 2523: 3036–3044.
3. Putzier M, Schneider SV, Funk JF, *et al*. The surgical treatment of the
   lumbar disc prolapse-nucleotomy with additional transpedicular dynamic
   stabilization versus nucleotomy alone. *Spine*, 2005, 30(5): E109–E114.
4. Madigan L, Vaccaro AR, Spector LR, *et al*. Management of sympto-
   matic lumbar degenerative disk disease. *J Am Acad Orthop Surg*, 2009,
   17(2): 102–111.

5. Hsu WK, Wang JC. The use of bone morphogenetic protein in spine fusion. *Spine J*, 2008, 8(3): 419–425.

6. Sharp CA, Roberts S, Evans H, *et al*. Disc cell clusters in pathological human intervertebral discs are associated with increased stress protein immunostaining. *Eur Spine J*, 2009, 18(11): 1587–1594.

7. Richardson SM, Hoyland JA. Stem cell regeneration of degenerated intervertebral discs: Current status. *Curr Pain Headache Rep*, 2008, 12(2): 83–88.

8. Beattie PF. Current understanding of lumbar intervertebral disc degeneration: A review with emphasis upon etiology, pathophysiology, and lumbar magnetic resonance imaging findings. *J Orthop Sports Phys Ther*, 2008, 38(6): 329–340.

9. Hohaus C, Ganey TM, Minkus Y, *et al*. Cell transplantation in lumbar spine disc degeneration disease. *Eur Spine J*, 2008, 17(Suppl 4): 492–503.

10. Gruber HE, Johnson TL, Leslie K, *et al*. Autologous intervertebral disc cell implantation: A model using Psammomys obesus, the sand rat. *Spine*, 2002, 27(15): 1626–1633.

11. Nomura T, Mochida J, Okuma M, *et al*. Nucleus pulposus allograft retards intervertebral disc degeneration. *Clin Orthop Relat Res*, 2001, 389: 94–101.

12. Hiyama A, Mochida J, Iwashina T, *et al*. Transplantation of mesenchymal stem cells in a canine disc degeneration model. *J Orthop Res*, 2008, 26(5): 589–600.

13. Hoogendoorn RJ, Lu ZF, Kroeze RJ, *et al*. Adipose stem cells for intervertebral disc regeneration: Current status and concepts for the future. *J Cell Mol Med*, 2008, 12(6): 2205–2216.

14. Lu ZF, Doulabi BZ, Wuisman PI, *et al*. Differentiation of adipose stem cells by nucleus pulposus cells: Configuration effect. *Biochem Biophys Res Commun*, 2007, 359(4): 991–996.

15. Gaetani P, Torre ML, Klinger M, *et al*. Adipose-derived stem cell therapy for intervertebral disc regeneration: An *in vitro* reconstructed tissue in alginate capsules. *Tissue Eng Part*, 2008, 14(8): 1415–1423.

16. Hoogendoorn R, Doulabi BZ, Huang CL, *et al*. Molecular changes in the degenerated goat intervertebral disc. *Spine*, 2008, 33(16): 1714–1721.

17. Tapp H, Hanley EN, Patt JC, *et al*. Adipose-derived stem cells: Characterization and current application in orthopaedic tissue repair. *Exp Biol Med*, 2009, 234(1): 1–9.

18. Kandel R, Roberts S, Urban JP. Tissue engineering and the intervertebral disc: The challenges. *Eur Spine J*, 2008, 17(Suppl 4): 480–491.
19. Halloran DO, Grad S, Stoddart M, *et al.* An injectable crosslinked scaffold for nucleus pulposus regeneration. *Biomaterials*, 2008, 29(4): 438–447.
20. O'Halloran DM, Pandit AS. Tissue-engineering approach to regenerating the intervertebral disc. *Tissue Eng*, 2007, 13(8): 1927–1954.
21. Boyd LM, Carter AJ. Injectable biomaterials and vertebral endplate treatment for repair and regeneration of the intervertebral disc. *Eur Spine J*, 2006, 15(Suppl 3): S414–S421.
22. Seguin CA, Grynpas MD, Pilliar RM, *et al.* Tissue engineered nucleus pulposus tissue formed on a porous calcium polyphosphate substrate. *Spine*, 2004, 29(12): 1299–1306.
23. Mizuno H, Roy AK, Zaporojan V, *et al.* Biomechanical and biochemical characterization of composite tissue-engineered intervertebral discs. *Biomaterials*, 2006, 27(3): 362–370.
24. Huang B, Li CQ, Zhou Y, *et al.* Collagen II/hyaluronan/chondroitin-6-sulfate tri-copolymer scaffold for nucleus pulposus tissue engineering. *J Biomed Mater Res B Appl Biomater*, 2010, 92(2): 322–331.
25. Richardson SM, Hughes N, Hunt JA, *et al.* Human mesenchymal stem cell differentiation to NP-like cells in chitosan glycerophosphate hydrogels. *Biomaterials*, 2008, 29(1): 85–93.
26. Edlund U, Danmark S, Albertsson AC. A strategy for the covalent functionalization of resorbable polymers with heparin and osteoinductive growth factor. *Biomacromolecules*, 2008, 9(3): 901–905.
27. Li J, Zhu B, Shao Y, *et al.* Construction of anticoagulant poly(lactic acid) films via surface covalent graft of heparin-carrying microcapsules. *Colloids Surf B Biointerfaces*, 2009, 70(1): 15–19.
28. Clouet J, Vinatier C, Merceron C, *et al.* The intervertebral disc: From pathophysiology to tissue engineering. *Joint Bone Spine*, 2009, 76(6): 614–618.
29. Masuda K. Biological repair of the degenerated intervertebral disc by the injection of growth factors. *Eur Spine J*, 2008, 17: S441–S451.
30. Masuda K, Imai Y, Okuma M, *et al.* Osteogenic protein-1 injection into a degenerated disc induces the restoration of disc height and structural changes in the rabbit anular puncture model. *Spine (Phila Pa 1976)*, 2006, 31(7): 742–754.
31. Thompson JP, Oegema TJ, Bradford DS. Stimulation of mature canine intervertebral disc by growth factors. *Spine*, 1991, 16(3): 253–260.

32. Kim DJ, Moon SH, Kim H, *et al.* Bone morphogenetic protein-2 facilitates expression of chondrogenic, not osteogenic, phenotype of human intervertebral disc cells. *Spine*, 2003, 28(24): 2679–2684.
33. Modric J. Bulging and herniated disc in the neck and lower back. Published December 17, 2015. <https://www.ehealthstar.com/conditions/bulging-herniated-disc>
34. Tintle S, Gwinn D. Intervertebral disc process of degeneration: Physiology and pathophysiology. Published February 4, 2015. <https://clinicalgate.com/intervertebral-disc-process-of-degeneration-physiology-and-pathophysiology/>

# Chapter 4

# Cell and Molecular Therapy in the Study of Intervertebral Disc Degeneration

The main function of the disc is to increase the flexibility of the vertebral body. The intervertebral disc absorbs the axial force of the vertebral body, promotes the transmission of the load, and allows the spine to move in a multiaxial direction. The intervertebral disc consists of various components such as the internal nucleus pulposus (NP) and outer annulus fibrosus (AF), which are responsible for composition and connection of the cartilage of the vertebra. A pathological state develops when the microenvironment of the intervertebral disc changes. Changes in the phenotypic and chemical factors of intervertebral cells will exacerbate cell disorder or accelerate intervertebral disc degeneration; the microenvironment of metabolic disorder is often accompanied by inflammation which, in a vicious spiral, further aggravates the condition. Tissue fibrosis and tissue dehydration lead to a decrease in proteoglycan content in the nucleus pulposus mainly due to loss of intervertebral height [1–3]. There may be rifts in the weak annulus, leading to further expansion of the nucleus pulposus. If the fibrous ring ruptures, the nucleus pulposus may come out of the intervertebral disc. Intervertebral instability, loss of intervertebral height, nucleus pulposus prolapse, or adjacent nerve root involvement are the main factors causing low back pain and leg pain. These are usually considered to be related to degeneration induced by genetic susceptibility and environmental factors, such as improper weight, lack of nutrition, obesity, and other diseases. Genetic factors also play an important role. Many studies show that gene alterations play an important role in protein metabolism and cell arrangement, which can accelerate the degeneration of the intervertebral disc [4, 5]. There is an inseparable relationship between degeneration of intervertebral disc and age because chronic disc degeneration is part of the natural aging process of the human body. However, the effects of one or more factors can exacerbate denaturation. Therefore, it is key that treatment is provided according to the different causes of the disease, which can provide the basis for individualized treatment. At present, the main methods in this area are the use of progenitor cells and cytokines [6].

## 1. Microenvironment of the Intervertebral Disc

The complex anatomical mechanism of the intervertebral disc provides good activity and shock absorption. The regularly arranged annulus fibrosus, nucleus pulposus, and cartilage plate constitute a unique, relatively independent microenvironment. The intervertebral disc is the largest non-vascular tissue in the human body, which is mainly nourished by the trophoblast system. The peripheral capillaries can only reach the outermost millimeter of the annulus fibrosus. Nutrients and metabolites can only enter human cartilaginous plates and fibrous rings through diffusion and permeation. As this region is far away from the blood vessel, the internal nucleus lacks a proper nutrient supply system, and thus this is a fragile, anaerobic, and low pH environment, which therefore affects its survival ability. Aging leads to cartilage calcification, and the nucleus pulposus cells further reduce the viability of fiber ring disc cells. Although the living environment is very fragile, some studies suggest that certain nucleus pulposus cells can still change the microenvironment. The effects of hypoxia-inducible factor (HIF) in the apoptosis of nucleus pulposus cells play an important role in changing the microenvironment. HIF belongs to the basic helix loop helix (BHLH) family, PERARNT SIM subfamily, which can affect cellular oxygen metabolism, especially in an anoxic environment. It is known that HIF-1, a part of the HIF family, is an important factor in cell energy metabolism and survival. HIF-1 is a heterologous dimer mainly formed by the HIF-1 instrument and HIF-1 beta two subunit. The stable expression of the subunit in cells plays a structural role; the regulation of the hypoxia signaling gene leads to HIF-1 activity [7]. Usually, HIF-1 changes the regulation of mitochondrial metabolism. Although the direct cause is not clear, it may be related to the anaerobic metabolism of the nucleus pulposus, because its energy metabolism is mainly dependent on glycolysis and glucose transport of the HIF target gene.

In addition to cellular oxygen levels, the axial load metabolism, shear force, bending, and biomechanics of the thoracic spinal cord, can affect the microenvironment of the intervertebral disc, especially the permeability of nucleus pulposus cells in resistance to axial load pressure, where the pressure is higher than that in normal tissues of 200 mOsm/kg. This shows that the transcription factor TonEBP NFAT5/ plays a role in adjusting nonionic osmotic pressure level, with the help of such compounds as betaine, taurine, inositol, and sorbitol, to help maintain the intervertebral disc in a stable state along with the hypertonic solution. Also, single cells, cell proliferation, low pH, high mucopolysaccharide level, and other factors can affect the microenvironment of the intervertebral disc [8–11]. A large number of studies have shown that the specific micro-environment in the intervertebral disc is indispensable, but the most important aspect is how to apply it to the treatment techniques so that the microenvironment of the intervertebral disc is retained.

## 2. Cell Therapy

### 2.1. *Intervertebral disc cell transplantation*

In the past twenty years, autologous chondrocyte transplantation for the repair of articular cartilage has become a mature treatment method, but there are some difficulties when this method is applied to autologous intervertebral disc cell transplantation because the intervertebral disc cells from normal tissues can lead to donor-onset intervertebral disc instability and accelerate the degeneration of the intervertebral disc. Recently, research on low-temperature storage of nucleus pulposus cells has confirmed that cell proliferation, matrix production, and storage of freshly isolated cells lead to no obvious change; surgery and activation of transplanted cells can destroy primitive cells, but this is not always in line with the needs of the patients with transplanted cells. It is also possible to choose an allograft made of allogeneic cells. In fact, in clinical and imaging studies, it is confirmed that allograft transplantation of immature chondrocytes combined with protein carrier in the treatment of moderate disc degeneration can decrease the degree of lesions.

### 2.2. *Mesenchymal stem cell transplantation and regeneration mechanism*

Mesenchymal stem cells (MSCs) have been selectively differentiated into intervertebral disc cells or chondrocytes [12–15]. They have been extensively studied in animal models of intervertebral disc degeneration and in some clinical trials. Although adipose tissue can separate mesenchymal stem cells and show the potential for regeneration in the intervertebral disc, bone marrow is still the most common source of stem cells. The mechanism of transplantation of mesenchymal cells for the treatment of degenerated intervertebral disc is still not completely clear. It has been observed that by co-culture or transplantation of nucleus pulposus cells to the intervertebral disc tissues, the mesenchymal stem cells could differentiate into nucleus pulposus-like cells. Studies have found that mesenchymal stem cells are injected into intervertebral discs of rats, rabbits and dogs. It is confirmed that mesenchymal stem cells can differentiate into intervertebral disc's corresponding cartilage or intervertebral disc-like cells. Liu *et al.* found that owing to the similar morphology and phenotype of nucleus pulposus of cartilage cells, cartilage can be formed, and they concluded this by assessing marker proteins, protein polysaccharides, and type II collagen production to evaluate the intervertebral disc-like cell differentiation. It is therefore necessary to find a special marker for identifying the intervertebral disc cells, which can be identified in the induced mesenchymal stem cells [16]. However, it is still not possible to determine whether mesenchymal stem cells separated by bone marrow and adipose tissue have the ability to differentiate into nucleus pulposus cells. In addition, the effect of the extracellular matrix produced by these induced cells on the repair of intervertebral disc is not clear. Again, there is no conclusive statistical data about the survival rate of mesenchymal stem cells transplanted into the intervertebral disc tissue in patients with hypoxia, low sugar, acidity, hyperosmotic condition, and in complex mechanical loading environments. It has been reported that the injected cells may survive for up to six months in the intervertebral disc tissue. Although MSC has an ideal outcome in the treatment of intervertebral disc degeneration, there are still many shortcomings and risks in promoting disc

regeneration through the injection of MSCs, such as cell leakage, osteoblast differentiation, and so on [17–19]. Therefore,
the standard flow for MSC collection, processing, and culture
conditions needs to be further improved so that MSC can play
a role in the treatment of degenerative disc degeneration.

## 2.3. *Homing of progenitor cells*

In order to avoid the problems and risks associated with cell
transplantation, the migration of regenerated cells to the site of
damage can be studied as a new research concept. In the process of natural restoration, the affected cells and tissues release
signals to attract the progenitor cells from their niche to the
injured site. These special chemical inducers include growth
factors, cytokines, and chemokines, the specific ones released
depending on tissue repair. In the *in vitro* mode, human
BMSCs can penetrate the endplate of the intervertebral disc to
repair the damaged tissue. Under the condition of induced
intervertebral disc degeneration, such as the restriction of
nutrition factors and the high load state or due to a puncture
needle, the old nest of the cell can be improved obviously
[20–23]. Therefore, selective therapy can increase chemical
chemotaxis to promote the homing of MSC. Future studies
should further assess the potential and safety of the chemical
inducer transmission system.

## 2.4. *Endogenous stem cells of the*
   *intervertebral disc*

Although a new treatment can be provided based on the
homing mechanism of the progenitor cells, the number of
cells available in NP is limited due to the absence of blood
vessels in the intervertebral disc. Therefore, it is very important to explore the feasibility and limitations of endogenous

self-repair. A recent study explored the presence of endogenous progenitor cells in the intervertebral disc for the repair of the intervertebral disc [24–27]. The researchers used the rabbit stem cells calibrated by 5-bromo-2'-deoxyuridine to determine the cell proliferation area and found that the cells proliferated slowly in NP and AF. The study showed that autologous (Notchl, Delta4, Jaggedl, C-KIT, K167, Stro-l) labeled progenitor cells were detected in the intervertebral disc in rabbits, rats, pigs, and humans. Many marker-positive cells appear in the corresponding regions, such as the AF external peripheral ligament and cartilage plate, indicating the existence of stem cells in the intervertebral disc microenvironment. These studies indicate that stem cells of specific tissues can improve their self-repair ability.

Studies on intervertebral disc progenitor cells have been carried out for many years. Earlier studies focused on fusion line extraction discs in cervical spondylosis patients whose intervertebral disc tissues were extracted, showing the presence of many factors, e.g. CD105, CD166, CD63, CD49a, D90, CD73, p75 and CD133/1 nerve growth factor, and a bone marrow mesenchymal stem cell phenotype. Next, *in vitro* growth of the bone marrow mesenchymal stem cells, and the role of osteoblasts, chondrocytes, and adipocytes, and differentiation in human disc tissue culture degeneration were studied. Further research showed that there exist stem cells with MSC tube implantation of gang culture isolated from patients with NP in the 16-row disc of human AF cells and NP cells. Compared with the extraction from the bone marrow of MSC differentiation, the results show that the MSC marker NP extraction can occur when it is positive to fat cells, which shows a weak trend of differentiation. When the cells extracted from the cartilage plate are compared with the BMSCs from the same patient *in vitro*, it is found that cells from the cartilage plate are similar to the MSCs in cell morphology, proliferation rate, cell cycle, and immune and gene phenotypes

[28]. MSCs extracted from cartilage can be induced to differentiate into osteoblasts, adipocytes, and chondrocytes, and MSCs in chondrocytes and osteoblasts are stronger than in bone marrow mesenchymal stem cells. Although there is increasing evidence of the presence of progenitor cells in the intervertebral disc, there is no way to induce the differentiation of these multi-differentiated MSCs in a specific direction.

## 3.  Bioactive Factors or Genes

### 3.1. *Injecting protein or small molecule substance*

Human bioactive factors can activate or mobilize endogenous cells to reconstruct early-disordered intervertebral microenvironment faster than the transplantation of autologous or allograft cells. New studies have included the intradiscal injection of various growth factors such as OP, BMP-7, and GDF1 or a 5-group injection of small molecules such as simvastatin groups; in addition, platelet-rich plasma (PRP) or platelets contain many growth factors and protein products that play roles in hybrid bacteriolytic therapy and so can enhance cell growth and tissue repair. Some *in vivo* and *in vitro* experiments have shown that PRP can help in the regeneration of intervertebral disc cells. But considering the risk of inflammation and problems of angiogenesis and pathogenesis, the safety and effectiveness of PRP needs to be further studied.

### 3.2. *Gene therapy*

The advantage of gene therapy in comparison with human protein is that it can have a long impact. BMPs, CDF-5, LMP-I, Sox 9, ILra, and TIMP-1 have been used for gene transfection in intervertebral discs by two main methods: virus and non virus. Most of the experiments were conducted using adenovirus vectors. Gilbertson advocated the use of BMP-2 gene therapy.

Another study found that GDF-5 or BMP transfected by adenovirus enhanced human and bovine intervertebral disc cell proliferation with LMP-1, which can improve the repairing ability of the degenerative intervertebral disc in animal models. It was found that the height, MRI signal, gene expression, and protein expression of intervertebral disc were significantly improved after transfection of OP-1-1 and SOX-9 genes using adenoviral vectors in a rabbit degenerative model.

### 3.3. *Carrier system*

Carrier materials are often used to transport exogenous proteins, genes, and small molecules into human cells. Carriers can play a protective role in reducing leakage and prolonging the time of action. In the degenerative disc model of mice, rhGDF 5 and polylactic co-glycolic acid (PLGA acid) could be detected after a minimum of 42 days. Similar studies have found that anti-IL1's use of PLGA as a carrier can also continue to perform its function in the intervertebral disc.

## 4. Intracellular Signaling Pathway

Recently, the study of the specific role of signal pathways in intervertebral discs has attracted increased attention. Shima Kiwachimin's study on the microenvironment demonstrated that factors such as oxygen concentration, mechanical stress, and related inflammation can be regulated by various signaling pathways activated by cytokines and growth factors.

### 4.1. *NF-κB and MAPK pathway*

The main signaling pathways include NF-κB and B-MAPK, which are responsible for the regulation of proinflammatory mediators such as TINF, IL-1, or IL-6; these two pathways have been identified in several musculoskeletal diseases in which the inhibition of NF-κB pathway and the catabolism of the main regulator have been shown to be therapeutic targets for intervertebral disc herniation and related pain. When NF-κB induced oligodeoxynucleotide in rat models of intervertebral disc herniation, mechanical tenderness and heat hyperalgesia in mice were significantly inhibited, indicating that there is a potential therapeutic potential. The MAPK pathway is also identified in the intervertebral disc as a target for the control of anti-inflammatory and anticatabolic metabolism. The researchers investigated the effect of curcumin on the intervertebral disc; when using the IL-1 treatment plan for intervertebral disc cells in which degeneration was induced by inflammatory/catabolic cascade reaction, addition of curcumin was shown to significantly reduce IL-6, MMP-1, and MMP-13 levels, while p38 and ERK were activated. The pathway analysis showed that the expression of TLR-2 and down-regulation of MAPK kinase JNK inhibits stroke [29]. The active osmotic regulation of MAPK, the synthesis of matrix, the expression of integrin and the regulation of HIF-1 play an important role in the survival of NP cells.

## 4.2. *Wnt pathway, Notch pathway*

The Wnt pathway is also considered to be involved in the occurrence and development of intervertebral disc diseases. According to the classical groups, Wnt is a signal in classical pathway Wnt/β-catenin pathway and non-canonical pathway Wnt/polarity pathway; the latter can be further divided into cell polarity pathway and $Ca^{2+}$ pathway, but regulation of intervertebral disc cells of classical Wnt is not the only way the Wnt signal pathway functions. In a non-classical Wnt pathway, the protein kinase C (PKC) signal simulation led to excitation of propylene glycol monomethyl ether acetate

(PMA), and nucleus pulposus cells showed mRNA and a decrease in the protein levels of catenin. PMA can induce cell proliferation, cell cycle process, and polyproteoglycan-4 expression. There is no TCH signal transduction pathway in human intervertebral disc cells. The proliferation of progenitor cells is mediated by the hypoxia-sensitive Notch signal transduction pathway that regulates the Notch signaling part in the intervertebral disc cells and is also activated by hypoxia. A Notch signal transduction inhibitor, L685458, inhibits the induced hypoxia and reduces the AF cell proliferation via its actions on the luciferase gene, 12xCSL, and CBFI. When the degenerated intervertebral disc tissue was compared with the non-degenerated intervertebral disc tissue, the expression of Notch2 and IL significantly increased the TCH pathway in the vertebra.

# References

1. Le Maitre CL, Pockert A, Buttle. DJ, *et al*. Matrix synthesis and degradation in human intervertebral disc degeneration. *Biochem Soc Transac*, 2007, 35(4): 652–655.
2. Ito K, Creemers L. Mechanisms of intervertebral disk degeneration/ injury and pain: A review. *Global Spine J*, 2013, 3(3): 145–152.
3. Crunhagen T, Shirazi-Adl A, Fairbank JCT, *et al*. Intervertebral disk nutrition: A review of factors influencing concentrations of nutrients and metabolites. *Orthopedic*. Clinics of North America (2011), Oct;42(4): 465–477.
4. Urban JPG, Smith S, Fairbank JCT. Nutrition of the intervertebral disc. *Spine*, 2004, 29(23): 2700–2709.
5. Risbud MV, Schipani E, Shapiro IM. Hypoxic, regulation of nucleus pulposus cell survival: From niche to notch. *Am J Pathol*, 2010, 176(4): 1577–1583.
6. Tran CM, Shapiro IM, Risbud MV. Molecular regulation of CCN2 in the intervertebral disc: Lessons learned from other connective, tissues. *Matrix Biol*, 2013, 32(6): 298–306.
7. Agrawal A, Guttapalli A, Narayan S, *et al*. Norrnoxic stabilization of HIF-Iα drives glycolytic metabolism and regulates aggrecan gene

expression in nucleus pulposus cells of the rat intervertebral disk. *Am J Physiol-Cell Physiol*, 2007 Aug;293(2): C621–C631.

7a. Cajghate S, Hiyama A, Shah M, *et al.* Osmolarity and intracellular calcium regulate aquaporin2 expression through TonEBP in nucleus pulposus cells of the intervertebral disc. *J Bone Miner Res*, 2009, 24(6): 992–1001.

8. Xi Y, Kong J, Liu Y, *et al.* Minimally invasive induction of an early lumbar disc degeneration model in rhesus monkeys. *Spine*, 2013, 38: E579–E586.

9. Power KA, Grad S, Rutges JPHJ, *et al.* Identification of cell surface-specific markers to target human nucleus pulposus cells: Expression of carbonic anhydrase '(XII varies with age and degeneration. *Arthritis Rheumatism*, 2011, 63(12): 3876–3886.

10. Stoyanov JV, Cantenbein-Ritter B, Bertolo A, *et al.* Role of hypoxia and growth and differentiation factor-5 on differentiation of human mesenchymal stem cells towards intervertebral nucleus pulposus-like cells. *Eur Cell Mater*, 2011, 21: 533–547.

11. Henriksson HB, Svanvik T, Jonsson M, *et al.* Transplantation of human mesenchymal stems cells into intervertebral discs in a xenogeneic porcine, model. *Spine*, 2009, 34(2): 141–148.

12. Vanden Berg-Foels WS. *In situ* tissue regeneration: Chemoattractants for endogenous stem cell recruitment. *Tissue Eng Part B: Rev*, 2013, 20(1): 28–39.

13. Henriksson HB, Thornemo M, Karlsson C, *et al.* Identification of cell proliferation zones, progenitor cells and a potential stem cell niche in the intervertebral disc region: A study in four species. *Spine*, 2009, 34(21): 2278–2287.

14. Risbud MV, Guttapalli A, Tsain, *et al.* Evidence for skeletal progenitor cells in the degenerate human intervertebral disc. *Spine*, 2007, 32(23): 2537–2544.

15. Blanco JF, Graciani IF, Sanchez-Guijo FM, *et al.* Isolation and characterization of mesenchymal stromal cells from human degenerated nucleus pulposus: Comparison with bone marrow mesenchymal strornal cells from the same subjects. *Spine*, 2010, 35(26): 2259–2265.

16. Liu LT, Huang B, Li CQ, *et al.* Characteristics of stem cells derived from the degenerated human intervertebral disc cartilage endplate. *PLOS One*, 2011, 6: e26285.

17. Than KD, Rahman SU, Wang L, *et al.* Intradiscal injection of simvastatin results in radiologic, histologic, and genetic. Evidence of disc

regeneration in a rat model of degenerative disc disease. *Spine J*, 2014, 14(6): 1017–1028.

18. Pirvu TN, Schroeder JE, Peroglio M, *et al*. Platelet-rich plasma induces annulus fibrosus cell proliferation and matrix production. *Eur Spine J*, 2014, 23(4): 745–753.

19. Lemaitre CL, Hoyland JA, Freemont AJ. Interleukin-I receptor antagonist delivered directly and by gene therapy inhibits matrix degradation in the intact degenerate. Human intervertebral disc: An *in situ* zymographic and gene therapy study. *Arthritis Res Therapy*, 2007, 9(4): R83.

20. Gilbertson L, Ahn SH, Teng PN, *et al*. The effects of recombinant human bone morphogenetic protein-2, recombinant human bone morphogenetic protein-12, and adenoviral bone morphogenetic protein-12 on matrix synthesis in human annulus fibrosis and nucleus pulposus cells. *Spine J*, 2008, 8(3): 449–456.

21. Liang H, Ma SY, Feng G, *et al*. Therapeutic effects of adenovirus- mediated growth and differentiation factor-5 in a mice. Disc degeneration model induced by annulus needle puncture. *Spine J*, 2010, 100): 32–41.

22. Ren S, Liu Y, Ma J, *et al*. Treatment of rabbit intervertebral disc degeneration with co-transfection by adeno-associated virus-mediated SOX9 and osteogenic protein-I double genes *in vivo*. *Int J Molecular Med*, 2013, 32(5): 1063–1068.

23. Yan J, Yang S, Sun H, *et al*. Effects of releasing recombinant human growth and differentiation factor-5 from poly (lactic-co-glycolic acid) microspheres for repair of the rat degenerated intervertebral disc. *J Biomater Appl*, 2013, 29(1): 72–80.

24. Wuertz K, Vo N, Kletsas D, *et al*. Inflammatory and catabolic signalling in intervertebral discs: The roles of NF-κ B and MAP kinases. *Eur Cell Mater*, 2012, 23: 103–119.

25. Suzuki M, Inoue G, Cemba T, *et al*. Nuclear factor-kappa B decoy suppresses nerve injury and improves mechanical allodynia and thermal hyperalgesia in a rat lumbar disc herniation model. *Eur Spine J*, 2009, 18(7): 1001–1007.

26. Klawitter M, Quero L, Klasen J, *et al*. Curcuma DMSO extracts and curcumin exhibit an anti-inflammatory and anti-catabolic effect on human intervertebral disc cells, possibly by influencing TLR2 expression and JNK activity. *J Inflamm (Innd)*, 2012, 9(1): 29.

27. Kikuchi A, Kishida S, Yamamoto H. Regulation of Wnt signaling by protein–protein interaction and post-translational modifications. *Exp Molecular Med*, 2006, 38(1): 1–10.

28. Arai F, Hiyama A, Sakai D, *et al.* The expression and role of non-canonical (PKC) signaling in nucleus pulposus cell metabolism. *J Ortho Res*, 2012, 30(9): 1478–1485.

29. Silván U, Díez-Torre A, Arluzea J, *et al.* Hypoxia and pluripotency in embryonic and embryonal carcinoma stem cell biology. *Differentiation*, 2009, 78(2): 159–168.

# Chapter 5

# The Research Status and Progress of Stem Cell Therapy in the Regeneration of Degenerated Intervertebral Discs

Intervertebral disc degeneration is a genetically predisposed condition mediated by many factors such as the disease process, load, trauma, age, and infection, any or all of which can lead to intervertebral disc degeneration. Necrosis, cell aging, changes in the intracellular enzyme amount, and metabolic imbalance, could cause damage and loss of homeostasis of the intervertebral disc, which could result in herniation and degeneration, eventually causing nerve root and the spinal cord damage as a result of spinal stenosis. This can cause a range of clinical symptoms and signs (such as low back pain, lower limb paralysis, and pain) and can even lead to severe disability [1]. Although the treatment methods for intervertebral disc degeneration are continuously being improved, medicine and physiotherapy cannot treat the underlying etiology; surgery (such as total disc replacement) can help to solve the end-stage symptoms, but cannot reverse the disease progress, and can also change the spinal stress conduction direction [2], leading to new secondary complications. The aging of the global population has become increasingly prominent, and so the problem will consume the already limited medical resources; therefore, the use of stem cell technology to delay or reverse the disc degeneration has huge social and economic benefits.

# 1. Microenvironment of the Intervertebral Disc

The intervertebral disc is the body's largest hypovascular tissue under physiological conditions, the intervertebral space is relatively narrow and closed, there is a lack of blood supply and innervation, and it has the capability to withstand mechanical pressure, high osmotic pressure, and low oxygen conditions. In addition, the intervertebral disc can escape cellular and humoral immunity, so it is also immune privileged [2, 3].

The nucleus originates from the embryonic notochord, and the normal nucleus pulposus is made up of negatively charged proteoglycans and type II collagen, the protein composition of stent, and mucopolysaccharide-formed side chain due to molecular electronegative mutual exclusion. On scanning electron microscopy, a "brush"-like appearance was observed; these molecules with water maintained the nucleus pressure. Fiber ring is composed of 15–25 layers of bundled collagen fibers and elastic fibers coated with a protein polysaccharide; the outer annulus is mainly composed of type I collagen and fibroblast-like cells, and the inner ring is composed of type II collagen fibers, fiber cells, and cartilage cells. The end plate is similar in structure to the articular cartilage, which is composed of hyaluronic acid, linking protein, and type II collagen, and this plays an important role in regulating the delivery of nutrients in the intervertebral disc. The vertebral capillary distribution to the endplate edge disappeared, while the nutrients for the disc were mainly diffused through the capillary for transport to the nucleus; the cell metabolites were excreted into the extracellular matrix to spread into the bloodstream [4, 5]. The number of intervertebral disc cells is closely related to the distribution of the blood vessels in the endplate, but not directly related to the height of the intervertebral disc [6].

## 2. Histopathological Features of Intervertebral Disc Degeneration

There was a decrease in the number of nucleus pulposus cells during the intervertebral disc degeneration, and the nucleus pulposus was replaced by cartilage cells and extracellular matrix. With the degeneration of the intervertebral disc, the nucleus was dehydrated, highly reduced in size, and exposed to external pressure; the fiber ring structure was disordered (fracture, dislocation, bulging, and protruding); there was endplate bone thickening; and abnormal structures and localized bone fractures were observed. During the process of disc degeneration, the content of proteoglycan is reduced, degradation of dextran, chondroitin sulfate, and keratan sulfate are observed, reduction of the endplate moisture and type II collagen level is seen, and gradual transformation into mineralized tissue occurs, which is caused by occlusive capillaries in the endplate and nucleus pulposus tissue nutrient shortage. At the same time, the amount and activity of matrix metalloproteinase is further enhanced, accelerating the degeneration process. Degeneration of the intervertebral disc leads to loss of normal anatomical structure, abnormal spine movement, mechanical transmission shaft change and instability, nerve root and spinal cord compression, and adjacent intervertebral disc herniation, all of which cause a series of symptoms and signs [4, 5, 7].

## 3. Characteristics of Intervertebral Disc-derived Stem Cells

Disc degeneration begins with a decrease in cell function and cell number; the proliferation ability of intervertebral disc cells is poor, especially nucleus pulposus cells. The stem or progenitor cells play an important role in maintaining homeostasis in the organ tissue, and reduction of the number of stem cells or a decline in their activity is an important cause of organ function decline; thus to ensure steady intradiscal environment, there is a need to maintain the original number of intervertebral disc cells and cell activity. Risbud *et al.* [8] isolated $CD49a^+$, $CD75^+$, and $CD105^+$ stem cells for treatment of adult degenerative intervertebral disc tissue. Experiments showed that these cells have osteogenic, chondrogenic, and adipogenic differentiation potentials. Feng *et al.* [9] confirmed that the annulus progenitor cells expressed CD29, CD49a, CD166, and Strol protein, and when these cells were added into the culture medium to the fibrous ring, these primitive cells induced by different factors differentiated into bone, cartilage, nerve, and endothelial cells, showing that adult stem cells can serve as excellent seed cells for the treatment of degenerative intervertebral disc. Also, these cells are an ideal source of allograft seed cells for the repair of intervertebral disc degeneration.

## 4. Stem Cells from Other Sources

An *in vitro* study of intervertebral disc degeneration in which the stem cells were derived from other tissues also received extensive attention, and it was found that adipose-derived stem cells (ADSCs) can be used as an ideal source of stem cells. Chun *et al.* [10] used C-arm-guided puncture needle to create a rabbit lumbar disc degeneration model and observed the degeneration of lumbar intervertebral disc at different time points using MRI. Nineteen weeks later, the rabbit-derived ADSCs were injected into the degenerated intervertebral disc and safranin staining was performed after 10 weeks. After the ninth week, it was found that the intervertebral disc tissue began to degenerate, and had completely degenerated after 15 weeks. Ten weeks after the injection, the cells began to proliferate and differentiate, and the extracellular matrix and cartilage matrix content increased. Another study confirmed [11] that when rabbit SPIO-ADSCs were transplanted into rabbit intervertebral disc degeneration model, using the signal changes of MRI for the purpose of monitoring, the target signals in the implanted group compared with PBS injection group and blank control group were significantly increased, suggesting that ADSCs may help to slow down and repair degenerative intervertebral disc tissue. Ganey *et al.* [12] injected ADSCs into a dog to test the autologous degenerative intervertebral disc tissue. It was observed that the intervertebral disc morphology recovered and cell matrix synthesis increased; then ADSCs and hyaluronic acid as a mixture were injected into the corresponding intervertebral disc, which led to an increase in the intervertebral disc nucleus pulposus cell density and collagen type II synthesis. The protein test results confirmed that ADSCs can repair the degenerated intervertebral disc. Another study [13] confirmed that the culturing of ADSCs *in vitro* was poor but stable when grown in serum obtained from different manufacturers and in different batch cultures, and when using different differentiation factors in the culture medium, multilineage differentiation potential was observed; the ratio of ADSCs to bone marrow-derived mesenchymal stem cells (BMSCs) is high, and hence the source is considered to be widely available and the cells are also safe to harvest. The proliferation of ADSCs is less, but the two have similar biological characteristics, and so they can be applied for the treatment of degenerative intervertebral disc tissue as these are an ideal source of seed cells.

A study on GFP-Bcl-2 transfection of BMSCs [14] confirmed the expression of more proteoglycan and type II collagen than the non-transfected BMSCs — the expression is 6.2 times higher, indicating that transgenic BMSCs can differentiate into nucleus pulposus-like cells, thereby providing strong evidence for stem cell transplantation in the treatment of intervertebral disc degeneration. Korecki *et al.* [15] believe that mesenchymal stem cells co-cultured with notochordal cells can differentiate into nucleus pulposus cells and the fibrous ring cells at the same time; when mesenchymal stem cells and chondrocytes were cultured as the control group, the amount of the synthesized proteoglycan, laminin, and collagen increased significantly. Yet another study [16] demonstrated that human umbilical cord blood mesenchymal stem cells (hUCB-MSCs) transplanted into rabbit intervertebral disc tissue within 4 weeks (as assessed by fluorescence microscopy to analyze cell survival and proliferation) showed the phenotype of cartilage cells. An immunohistochemical test also confirmed these results and demonstrated that the type II collagen protein and polysaccharide levels in the target area increased. These results support the use of hUCB-MSCs for the repair of degenerative intervertebral disc cells.

A domestic study [17] has confirmed that a co-culture of notochordal cells and BMSCs for treatment of intervertebral disc cells, through mutual stimulation, can promote the proliferation and differentiation of BMSCs and can induce the production of chitosan and collagen type II, demonstrating the phenotype of chondrocytes. Chen *et al.* [18] co-cultured synovial stem cells and fibrous ring cells with different doses of transforming growth factor beta (TGF-β); a large majority of synovial stem cells showed differentiation potential to transform into nucleus pulposus cells, due to the synthesis of type II collagen and proteoglycan and increase in the level of SOX-9 protein. The *in vitro* co-culture provides a theoretical basis for the repair of intervertebral disc degeneration.

Another study [19] by using the co-culture method activated fibrous ring cells transplanted into the rabbit intervertebral disc degeneration model, and the degree of intervertebral disc degeneration was evaluated according to Nishimura's method at different time points. The normal group showed no obvious change; the score for the group where cells were not implanted in an intervertebral disc degeneration model was 2.8; in the group with the total implantation of mesenchymal stem cells and activation of annulus fibrosus cells, the score was 2.2; and in the group with the injection of mesenchymal stem cells and fibrous ring cells, the score was 1.2. These data confirm that mesenchymal stem cells can activate cells and annulus fibrosus cells in a co-culture. Watanabe *et al.* [19] co-cultured the fiber ring cells and mesenchymal stem cells, and the proliferation and differentiation capacity of the fiber ring cells and the increase in the amount of matrix synthesis were observed. Yamamoto *et al.* [20] co-cultured fibrous ring cells and mesenchymal stem cells in a biofilm system with a 0.45 m pore diameter to separate the two, but when observed by scanning electron microscopy it was seen that the two kinds of cells were significantly in contact with each other. DNA and protein polysaccharides were detected by (3H)glucosamine and (35S)sulfate, respectively, and the presence of various growth factors in the cell culture supernatant was detected. The amount of DNA and protein synthesized by polysaccharide annulus cells increased, and in the supernatant, the levels of TGF-beta 1, platelet-derived growth factor (PDGF), insulin-like growth factor (IGF-1), and other cytokines increased significantly. A probable explanation for this is that the cells through cell–cell contact increase the secretion of different cell growth factors to stimulate the proliferation and differentiation and interaction. Strassburg *et al.* [21]

used STraS cells and mesenchymal stem cells to assess the aging of the fiber ring cells and found that this could stimulate the differentiation of mesenchymal stem cells into fiber ring-like cells while leading to an increase in extracellular matrix synthesis.

## 5. Transplantation of Stem Cells into the Intervertebral Disc *In Vivo*

Sakai *et al.* [22] studied stem cells transplanted into rabbit degenerated intervertebral disc tissue by transplantation with green fluorescent protein-transfected stem cells. Upon observation after 48 weeks with MRI, results confirmed the survival of the implanted cells in the nucleus and the signal intensity of degenerative intervertebral disc increased. Immunohistochemical results confirmed that the target cells successfully transplanted into the degenerative intervertebral disc tissue, also survival, proliferation and differentiation into nucleus pulposus cells were confirmed. A study [13] with hyaluronic acid as a carrier and 15% of mesenchymal stem cells carrying green fluorescent protein were injected into rat caudal disc; 4 weeks later, cell number, viability, and the corresponding intervertebral disc height were maintained and mesenchymal stem cell proliferation in the intervertebral disc tissue was observed. A study [23] on co-cultured MSCs and nucleus pulposus cells with fibrin glue for rat tail disc carrier transplant assessed the results on days 14 and 35 after the rats were sacrificed. The intradiscal proteoglycan and cytokine content were detected, and it was observed that the levels were higher in the experimental group compared with the control group, and with time, the cells in the experimental group led to disc height being increased significantly. Hiyama *et al.* [24] injected mesenchymal stem cells into a dog intervertebral disc degeneration model, cell survival and proliferation were observed to be expressed similar to human nucleus pulposus cells and the fibrous ring of cell protein; immunohistochemistry showed that in the experimental group, compared with the control group, the expression of extracellular matrix was more and the target cells in the nucleus pulposus implant area also expressed FasL, thereby confirming that mesenchymal stem cells have similar cellular immunology and intervertebral disc cell changes.

Orozco *et al.* [25] studied 10 cases with complete fiber ring but with a diagnosis of lumbar degenerative disease in patients, after obtaining informed consent, in a clinical trial. The autologous BMSCs were transplanted into intervertebral disc lesions, and the patients were followed up for one year. The results showed that patients' pain significantly reduced and MRI showed a significant increase in the amount of water content in the lesion area, which is a significant biological effect compared with interbody fusion or intervertebral disc replacement surgery. Therefore, stem cell transplantation can be used to treat lumbar and back pain caused by chronic degenerative changes. Prologo *et al.* [26] radiolabeled human mesenchymal stem cells and under fluoroscopic guidance transplanted them into degenerated lumbar discs in pigs, and normal discs were used as the control group. The 3D implants were assessed by PET-CT scanning after 15 days to study the intervertebral disc tissue of the implanted area, and immunohistochemical staining was also done. The results after 3 days, as confirmed by PET-CT examination, showed that cells implanted into the intervertebral disc had survived and were proliferating; at 15 days, the extracellular matrix synthesis increased significantly. Yoshikawa *et al.* [27] studied two patients in whom autologous mesenchymal stem cells were injected into the disc vacuum zone, as confirmed by CT. In these two patients, the intervertebral disc "vacuum" area was significantly reduced, and T2W-MRI confirmed that disc signal cell transplantation area had increased.

## 6. Limitations and Prospects

Stem cells have been successfully used as treatment for diseases and tumors of the blood system, and now the technology is gradually being used to treat bone and joint diseases, but it is still in the early stages of research. The mechanism by which the stem cells transplanted into the relatively narrow and closed hypovascular denervated intervertebral disc tissue receive adequate nutrition and information regarding the direction of the proliferation and differentiation still needs further research. Vadal *et al.* [28] reported that the transplantation of stem cells for bone defects could lead to differentiation into tumor cells; so, for the technology to be successfully applied in the clinic, further study on this is required and care must be taken. However, with the continuous development of tissue engineering, genetics, and materials engineering, the use of stem cells for biological therapy, and even to reverse disc degeneration, will eventually become possible.

## References

1. Kepler CK, Ponnappan RK, Tannoury CA, *et al.* The molecular basis of intervertebral disc degeneration. *Spine J*, 2013, 13(3): 318–330.
2. Hughes SP, Freemont AJ, Hukins DW, *et al.* The pathogenesis of degeneration of the intervertebral disc and emerging therapies in the management of back pain. *J Bone Joint Surg Br*, 2012, 94(10): 1298–1304.
3. Huang S, Tam V, Cheung KM, *et al.* Stem cell-based approaches for intervertebral disc regeneration. *Cur Stem Cell Res Ther*, 2011, 6(4): 317–326.
4. Bergknut N, Smolders LA, Grinwis GC, *et al.* Intervertebral disc degeneration in the dog. Part 1: Anatomy and physiology of the intervertebral disc and characteristics of intervertebral disc degeneration. *Vet J*, 2013, 195(3): 282–291.
5. Bergknut N, Meij BP, Hagman R, *et al.* Intervertebral disc disease in dogs. Part 1: A new histological grading scheme for classification of intervertebral disc degeneration in dogs. *Vet J*, 2013, 195(2): 156–163.

6. Boubriak OA, Watson N, Sivan SS, *et al.* Factors regulating viable cell density in the intervertebral disc: Blood supply in relation to disc height. *J Anat*, 2013, 222(3): 341–348.

7. Bergknut N, Rutges JP, Kranenburg HJ, *et al.* The dog as an animal model for intervertebral disc degeneration? *Spine (Phila Pa 1976)*, 2012, 37(5): 351–358.

8. Risbud MV, Guttapalli A, Tsai TT, *et al.* Evidence for skeletal progenitor cells in the degenerate human intervertebral disc. *Spine (Phila Pa 1976)*, 2007, 32(23): 2537–2544.

9. Feng G, Yang X, Shang H, *et al.* Multipotential differentiation of human annulus fibrosus cells: An *in vitro* study. *J Bone Joint Surg Am*, 2010, 92(3): 675–685.

10. Chun HJ, Kim YS, Kim BK, *et al.* Transplantation of human adipose-derived stem cells in a rabbit model of traumatic degeneration of lumbar discs. *World Neurosurg*, 2012, 78(3–4): 364–371.

11. Jiang Xinhua, Chen Jianyu, Cai Zhaoxi, *et al.* MRI tracer. SPIO labeled fatty degeneration of the intervertebral disc stem cell transplantation. *China Clin Rehabil Tissue Eng Res*, 2011, 15(40): 7515–7519.

12. Ganey T, Hutton WC, Moseley T, *et al.* Intervertebral disc repair using adipose tissue-derived stem and regenerative cells: Experiments in a canine model. *Spine (Phila Pa 1976)*, 2009, 34(21): 2297–2304.

13. Jeong JH, Lee JH, Jin ES, *et al.* Regeneration of intervertebral discs in a rat disc degeneration model by implanted adipose-tissue-derived stromal cells. *Acta Neurochir (Wien)*, 2010, 152(10): 1771–1777.

14. Fang Z, Yang Q, Luo W, *et al.* Differentiation of GFP-Bcl-2-engineered mesenchymal stem cells towards a nucleus pulposus-like phenotype under hypoxia *in vitro. Biochem Biophys Res Commun*, 2013, 432(3): 444–450.

15. Korecki CL, Taboas JM, Tuan RS, *et al.* Notochordal cell conditioned medium stimulates mesenchymal stem cell differentiation toward a young nucleus pulposus phenotype. *Stem Cell Res Ther*, 2010, 1(2): 18.

16. Anderson DG, Markova D, An HS, *et al.* Human umbilical cord blood-derived mesenchymal stem cells in the cultured rabbit intervertebral disc: A novel cell source for disc repair. *Am J Phys Med Rehabil*, 2013, 92(5): 420–429.

17. Zhang Yannan, Shao Zengwu, Wu Yongchao, *et al.* Research of notochordal cells promote bone marrow mesenchymal stem cell proliferation and directional induced differentiation. *J Exp Surg*, 2012, 29(7): 1271–1274.

18. Chen S, Emery SE, Pei M. Coculture of synovium-derived stem cells and nucleus pulposus cells in serum-free defined medium with supplementation of transforming growth factor-beta1: A potential application of tissue-specific stem cells in disc regeneration. *Spine (Phila Pa 1976)*, 2009, 34(12): 1272–1280.

19. Watanabe T, Sakai D, Yamamoto Y, *et al.* Human nucleus pulposus cells significantly enhanced biological properties in a co-culture system with direct cell-to-cell contact with autologous mesenchymal stem cels. *J Orthop Res*, 2010, 28(5): 623–630.

20. Yamamoto Y, Mochida J, Sakai D, *et al.* Upregulation of the viability of nucleus pulposus cells by bone marrow-derived stromal cells: Significance of direct cell-to-cell contact in co-culture system. *Spine (Phila Pa 1976)*, 2004, 29(14): 1508–1514.

21. Strassburg S, Richardson SM, Freemont AJ, *et al.* Co-culture induces mesenchymal stem cell differentiation and modulation of the degenerate human nucleus pulposus cell phenotype. *Regen Med*, 2010, 5(5): 701–711.

22. Sakai D, Nakamura Y, Nakai T, *et al.* Exhaustion of nucleus pulposus progenitor cells with ageing and degeneration of the intervertebral disc. *Nat Commun*, 2012, 3: 1264.

23. Allon AA, Aurouer N, Yoo BB, *et al.* Structured co-culture of stem cells and disc cells prevent disc degeneration in a rat model. *Spine J*, 2010, 10(12): 1089–1097.

24. Hiyama A, Mochida J, Iwashina T, *et al.* Transplantation of mesenchymal stem cells in a canine disc degeneration model. *J Orthop Res*, 2008, 26(5): 589–600.

25. Orozco L, Soler R, Morera C, *et al.* Intervertebral disc repair by autologous mesenchymal bone marrow cells: A pilot study. *Transplantation*, 2011, 92(7): 822–828.

26. Prologo JD, Pirasteh A, Tenley N, *et al.* Percutaneous image-guided delivery for the transplantation of mesenchymal stem cells in the setting of degenerated intervertebral discs. *J Vasc Interv Radiol*, 2012, 23(8): 1084–1088.e6.

27. Yoshikawa T, Ueda Y, Miyazaki K, *et al.* Disc regeneration therapy using marrow mesenchymal cell transplantation: A report of two case studies. *Spine (Phila Pa 1976)*, 2010, 35(11): E475–E480.

28. Vadal G, Sowa G, Hubert M, *et al.* Mesenchymal stem cells injection in degenerated intervertebral disc: Cell leakage may induce osteophyte formation. *J Tissue Eng Regen Med*, 2012, 6(5): 348–355.

# Chapter 6

# Seed Cells

## 1. The Biological Properties and Requirements of Seed Cells

Seed cells for intervertebral disc tissue engineering should have the following characteristics: safe, effective, reliable, can be easily obtained with little damage to the body, ease of mass culture, and stable amplification and passaging, leading to stable phenotypes *in vitro*. They should also have lower immunogenicity, cancer risk, and genetic defects. The differentiation ability and rate of proliferation are reliable. They have similar biomechanical properties to the normal intervertebral disc *in vitro* and when cultured on scaffolds, and after *in vivo* implantation, show good biocompatibility.

## 2. Source of Seed Cells

At present, the main source of seed cells for tissue engineering of intervertebral discs includes autologous or allogeneic ring cells, nucleus pulposus cells, cartilage cells, and notochord-derived cells. Other sources include the bone marrow, muscle, blood vessels, nerves, and other types of cells of the spinal cord such as mesenchymal stem cells, embryonic stem cells, and genetically modified progenitor cells [1, 2].

## 2.1. *Intervertebral disc annulus, nucleus, and spinal cord-derived cells*

### 2.1.1. *Annulus fibrosus*

In anatomy, it refers to the outer periphery of a disc fiber structure arranged in concentric circles. The outer layer is mainly made up of collagen fibers, and the inner layer is a fibrous cartilage zone. In the intervertebral disc, there are some progenitor cells that differentiate and proliferate into fibers. Research has shown that these kinds of cells are similar to stem cells, autologous tissue cells, because there is no rejection and their safety is good when compared to xenogeneic cells; hence, these cells are used in intervertebral disc tissue engineering.

Feng [3] isolated fibrous ring cells from the human intervertebral disc annulus by flow cytometry, and using immunofluorescence detection showed that they expressed mesenchymal stem cell markers and cell surface antibodies such as CD29, CD49e, CD51, CD73, CD90, CD105, CD166, CD184, and Stro-1, thus demonstrating that they can differentiate into fat cells, osteoblasts, cartilage cells, fibroblasts, nerve cells, and endothelial cells. Saraiya [4] used rabbit intervertebral disc cells with purine analogs in a reverse osteogenic medium and in 1–4 days, due to the enhanced expression of alkaline phosphatase, osteocalcin, bone sialic acid glycoprotein, and collagen were formed. In an adipogenic induction medium, cytosolic lipid droplet accumulation increased, and PPAR-g2, LPL, and Fabp levels also increased, as assessed by the expression of mRNA, in the chondrogenic medium. All of this induced protein production also led to increase in the levels of chitosan, collagen II, IX, and XI, and versican. This has demonstrated that fibrous ring cells possess plasticity after reversal of hormone treatment, thus leading to their enhanced ability to undergo mesenchymal differentiation. Yang [5] grew fiber ring cells in alginate glass powder in a bioreactor, and the vessel was rotated 3 weeks after culture. Then, using real-time RT-PCR, the levels of aggrecan, collagen II, and collagen proline hydroxylase 4 were found to have increased significantly. Biochemical analysis shows that the levels of proteoglycan and collagen in dynamic cell culture are 2–5 times that seen in statically cultured cells; the corresponding DNA content is 3 times higher. After 1 week of culture, the proliferating cell nuclear antibody expression in fibrosus cells increased significantly in the bioreactor, thereby demonstrating that a bioreactor culture can be used in the field of tissue engineering for fiber ring cell amplification.

In the presence of fiber ring cells, progenitor cells, which can proliferate and differentiate into bone and cartilage *in vitro*, can be the potential source of seed cells. However, the number of such cells in the fiber ring is small, they are difficult to isolate and screen, and their *in vitro* expansion and growth are slow, all of which restricts their application in tissue engineering.

### 2.1.2. *Nucleus pulposus*

With aging, the cell components and water content decrease, there is a decrease in the synthesis of extracellular matrix components, and type I collagen is gradually replaced by collagen type II, but there is still a small amount of stem cells and progenitor cells that retain their differentiation ability, which can help to prevent intervertebral disc degeneration. So, these cells, to a certain extent, can be used as seed cells.

Henriksson [6] studied disc cell proliferation in normal intervertebral disc area, and also studied intervertebral disc degeneration using Notch1 antibody, Delta4 antibody, Jagged1 antibody, C-KIT antibody, KI167 antibody, and Stro-1 antibody in rabbit, rat, pigs, and found that the annulus and nucleus structure demonstrated higher expression of 5-bromo-2-deoxyuridylic, and these cells also showed low proliferation; in ligament and cartilage membrane area in the fiber ring boundary that possesses a stem cell niche-like structure, during degenerative intervertebral disc disease, the morphology and function of these cells may play a role in preventing disease progression. Blanco [7], from 16 cases of intervertebral disc degeneration, simultaneously collected intervertebral disc cells and iliac crest cells and on comparison found that amplification time, immunophenotype, differentiation, and molecular biology were similar to that observed in MSCs upon standard treatment following isolation from the nucleus progenitor cell. However, there was no differentiation to the fat cell lineage, and thus this study was different from a previous one. Rutges [8] studied gene and protein variation in human nucleus pulposus cells, fibrous ring cells, and cartilage cells and found that fibrous ring cells and cartilage cells had higher levels of cytokeratin 19 (KRT19) mRNA and neural cell adhesion molecule 1 compared to the nucleus pulposus cells; also increased expression of cartilage cell keratin 18 was observed. Cartilage oligomeric matrix protein (COMP) and osteocalcin (MGP) matrix and pleiotrophin expressions are higher in chondrocytes, and it has been found in nucleus pulposus cells that KRT19 level increases with age and then decreases, but MGP increases, thus leading to the hypothesis that the KRT19 gene expression level may be a characteristic of human NP cells. Minogue [9], using microarray technology, compared and identified 34 normal nucleus pulposus specimens and 49 degenerative intervertebral disc specimens. Then, using the qRT-PCR verification, it was found that 4 genes (*SNAP25*, *KRT8*, *KRT18*, and *CDH2*) showed significantly reduced expression in human degenerative nucleus pulposus cells and 3 genes (*VCAN*, *TNMD*, and *BASP1*) had significantly increased expression. *FBLN1* gene expression was found to be significantly decreased in the degenerative human annulus fibrosus and nucleus pulposus cells. Hegewald [10] in the study of protrusion of the intervertebral disc nucleus pulposus found that seed cells possessing the differentiation potential were obtained from the seed cells of nucleus pulposus for redifferentiation. From the nucleus of progenitor cells, the acquisition of cells is relatively easy, and the *in vitro* proliferation is fast, but there are also some problems, such as the changes in the biological characteristics seen with aging, as *in vitro* cultured cells when aging suffer from phenotypic instability, and thus, recently, their application in the study of tissue engineering has gradually reduced.

### 2.1.3. *Spinal cord-derived cells*

Spinal cord-derived cells (notochordal cell, referred to as NC), or notochord cells, exist in the nucleus, annulus, and endplate. Residual notochord is seen with development from birth, but begins to decrease from the age of 10, when it is replaced by cartilage cells. The same mechanism is seen after disc degeneration, but the mechanism is still not clear [11]. Erwin [12] obtained notochord cells from canine nucleus pulposus, cultured it in alginate beads as a monolayer and in three-dimensional space under hypoxic conditions (3.5% $O_2$) and in normal oxygen concentration (21% $O_2$). Histology, immunohistochemistry, and scanning electron microscopy were performed after 5 months, and the results showed that in the absence of oxygen (3.5% $O_2$), the notochord cell culture took the form of a similar composite and organized three-dimensional porous structure, rich in proteoglycan and collagen II. When cultured in normal oxygen concentration (21% $O_2$), the notochord cells failed to produce a similar structure. Therefore, while culturing these cells *in vitro*, attention should be paid to the oxygen conditions. Kim [13] used rabbit intervertebral disc nucleus pulposus cells for purification and screening followed by three-dimensional culture of notochord cells and chondrocytes. A comparison was then made of protein absorption of sulfate polysaccharide. In two-dimensional culture, immunohistochemical staining (collagen II, aggrecan, and SOX9) was performed, which demonstrated the ability of notochord cells to produce more proteoglycans than cartilage-like cells; also, the expression of collagen II, aggrecan, and SOX9 can help these cells differentiate into differentiated cartilage-like cells similar to the polymer. And hence they are used for the treatment of intervertebral disc degeneration as progenitor cells owing to their potential.

Notochord cells can inhibit the secretion of matrix and slow down intervertebral disc degeneration [14] to a certain extent, but there are very few notochordal cells in the intervertebral disc, thus making their separation, purification, screening, identification and stable amplification difficult. Also, the relationship between notochordal cells in the intervertebral disc nucleus pulposus and annulus and endplate cells is not completely clear. This poses a challenge for its application in tissue engineering, and once the abovementioned problems are solved, notochord cells are expected to become ideal seed cells.

## 2.2. *External source of intervertebral disc cells*

The seed cells from the external sources for intervertebral discs mainly include stem cells or progenitor cells with the ability of anisotropy or directional differentiation, such as embryonic stem cells, adult stem cells, and gene-modified progenitor cells.

### 2.2.1. *Embryonic stem cells*

Embryonic stem cells (ESC) can be derived from human or animal embryos or primordial germ cells and made to become a pluripotent cell line. They can undergo inward and outward differentiation of three types. Quintin [15] isolated stem cells from donor embryos and cultured them in alginate and the glass powder; after 28 days of culture they were found to produce proteoglycan and collagen II and expressed low levels of type I and X collagen, reflecting chondrogenic differentiation ability. The led to the theory that these can be used as seed cells for intervertebral disc degeneration repair. ESC has a strong ability of proliferation and differentiation, but its purification, differentiation, and regulation are difficult; tissue biocompatibility is not clear; long-term side effects and tumor risk still exist; and, especially, controversial ethical issues still surround the use of these cells. The abovementioned problems limit its wide application in the field of tissue engineering.

## 2.2.2. *Mesenchymal stem cells*

Mesenchymal stem cells (MSCs) can be of various types such as bone marrow mesenchymal stem cells (BMSCs), and they can also be seen in muscle, bone, cartilage, nerve and blood vessels, and other tissues and organs. Therefore, they are ideal for use as seed cells for intervertebral disc tissue engineering. BMSCs can differentiate into fibroblast-like cells, and this has been confirmed by previous studies [16, 17]. BMSCs not only have differentiation ability, they can also secrete a variety of growth factors and cytokines [18, 19]. MSCs and nucleus pulposus cells when co-cultured can enhance cell proliferation, DNA synthesis, and proteoglycan synthesis, and growth factors may also be secreted by the MSCs to promote the proliferation and differentiation of nucleus pulposus cells [20, 21].

Yang [22] used 54 New Zealand rabbits and by needle aspiration removed the nucleus pulposus to induce intervertebral disc degeneration. The animals were then randomly divided into three groups: model group, degeneration of pure fibrin gel transforming growth factor beta 1 (PFG-TGF-beta 1) implantation group, and mesenchymal stem cells pure fibrin gel transforming growth factor beta 1 (MSC-PFG-TGF-beta 1) implantation group. After 4, 8, and 12 weeks of treatment, X-ray, MRI, and histological examination were performed which revealed that implantation of MSCs inhibited the apoptosis of nucleus pulposus cells and reduced the loss of disc height ratio; the results show that the MSC-PFG–MRI TGF-beta 1 group relative to the other two groups revealed the least decrease in the height and minimum loss and an increase in the content of type II collagen in the nuclear marrow.

MSCs have low immunogenicity and undergo induced differentiation, can proliferate *in vitro*, and are relatively stable. There are more than 112 cases of clinical trials in patients with autologous chondrocyte transplantation for intervertebral disc treatments [23, 24]. The theory of using MSCs in cell therapy and tissue engineering has been studied and it has shown advantages (Fig. 1). But the practical application of MSCs still faces difficulties, such as the problem of *in vivo* safety after implantation, operability and final effect of the cells, the best number of cells that need to be implanted, etc. [25] *In vitro* culture techniques and additives may increase the risk of infection from homologous gene complex formation [26]; amplification may lead to accumulation

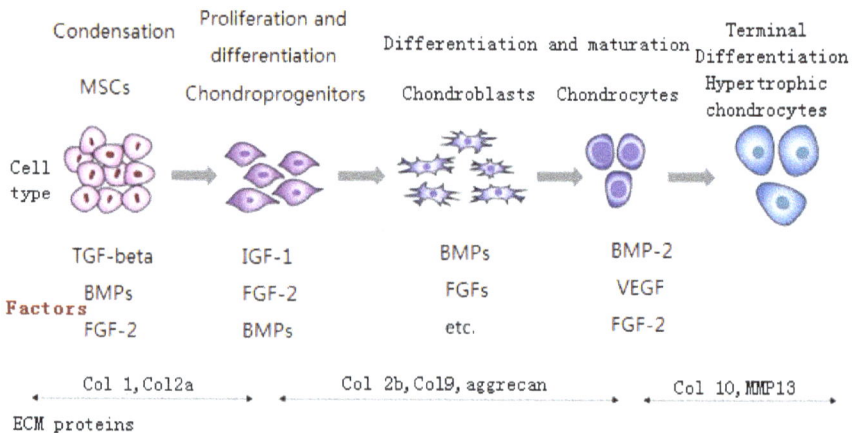

Fig. 1. MSCs undergo induced differentiation.

of genetic and epigenetic changes in the cells *in vitro*, which can lead to malignant changes [25, 26]. Chronic rejection of the implanted cells is also a possibility that may lead to shorter survival. Therefore, a long-term safety study is required [26].

### 2.2.3. *Other tissue-derived progenitor cells*

Recently, some researchers have isolated cells from the olfactory system, skin, nerves, muscles, tendons, and other tissues and organs and cultured them for proliferation and also differentiated them into adult adipose tissue, bone, synovium and synovial membrane, and trabecular bone [27–35]. Murrell [29] isolated stem cells from rabbit olfactory mucosa for *in vitro* proliferation and differentiation into chondrocytes for intervertebral disc treatment; the progenitor cells were implanted and *in vitro* and *in vivo* experiments showed that olfactory nerve-derived cells demonstrated intervertebral disc damage response signal. Goldschlager [36] amplified for isolation *in vitro* immune sheep bone marrow mesenchymal progenitor cells (mesenchymal progenitor cells, MPCs) and with the regulation by chondrogen/ poly-pentosan nitrate, the cell cage gelatin sponge was transformed into sheep cervical vertebrae (C3/4). After 3 months, according to the histological results, in the experimental group the intervertebral cartilage tissue increased significantly compared with the control group, which showed that MPCs combined with chondrogen/pentosan poly-nitrate can be used to produce cartilage tissue for intervertebral disc replacement. Sommar [37] obtained from healthy humans dermal fibroblasts and used a platelet-rich gel with macroporous microcarriers for an encapsulated cystic three-dimensional culture, and thus confirmed the *in vitro* formation of bone- or cartilage-like tissue in the induction medium under specific conditions. Also, Jackson used cells from human muscle tissue to obtain stromal progenitor cells, and compared these MSCs in terms of morphology, proliferation, cell surface, and differentiation ability and determined that they are very similar [28].

## 2.2.4. *Gene-modified progenitor cells*

Directly isolated and cultured progenitor cells or stem cells suffer from problems of slow growth, poor biocompatibility, and single seed cells and cannot meet the needs for ideal tissue engineering; hence, genetically modified stem cells or progenitor cells may be much more useful for specific applications owing to their adaptability, and thus they have become the focus of current research. Jiangbiao [38], using cationic liposome, mediated BMP-9 gene transfection of rabbit bone marrow mesenchymal stem cells; the gene transfection rate after 24 hours of transfection in the experimental group (34.45%) was significantly higher than that in the control group (0.40%). The ALP transfection group was assayed, and it was found that the activity in this group was significantly higher than that of the untransfected group. H&E staining of the scaffold material after implantation in rabbit muscle after 4–8 weeks indicated that island and banded cartilage were seen in the experiment group, whereas the bone matrix formation in the control group was only partially completed. This showed that BMP-9 gene-modified MSCs have better osteogenic ability compared with MSCs. Jiang [39] used lentiviral vectors hBMP2 and hVEGF165 and co-transfected genes into MSCs and, by enzyme-linked immunosorbent assay, confirmed the successful expression of two genes in MSCs. They found that 14 days later, the two gene-modified cells showed the highest alkaline phosphatase activity.

Induced pluripotent stem cells (iPSCs) possess four transcription factors (Oct3/4, Sox2, KLF4 and c-Myc) that confer to them the ability to differentiate and mature into somatic cells. Somatic cell-specific marker gene expression can be inhibited, and this leads to a specific phenotype of pluripotent cells, similar to what is obtained when using ESCs, and thus one can avoid the ethical issues and have a new way of providing seed cells for tissue repair [40]. A study by Galende [41] found that patient-derived iPSCs with genetic and immunological aspects can be used in tissue engineering research, but due to the introduction of foreign genes in iPSCs they carry a high tumor risk, and so the question of whether they can be used in intervertebral disc tissue engineering needs further study. At present, through genetic reprogramming means, the biological properties of stem cells can be adjusted, and so they give certain advantages in terms of the application prospects. But the research is still in the primary stage, and there are many technical problems that need to be overcome, such as the problem that addition of a homologous gene increases the risk of genetic defects, the high cancer risk, and the reprogramming control safety issues.

## 3. Summary

The research and application of tissue engineering requires a lot of seed cells, and it is difficult to find adequate, convenient sources of such cells with minimal trauma; there is also the problem of how to maintain fast proliferation and differentiation of these cells *in vitro*, since we know that a single cell cannot meet the requirements for tissue engineering. Korecki [42] found that notochordal cell-conditioned medium can promote MSCs to differentiate into cells with nucleus pulposus phenotype. Purmessur [43] used BMSCs and porcine notochordal cells in a co-culture and observed that on differentiation, notochord cells secreted soluble factors and promoted the differentiation of MSCs to nucleus pulpous cells, with high levels of synthesis of proteoglycan and collagen fiber, which could lead to hyperplasia inhibition.

In summary, fibrous ring cells have a stable phenotype, but suffer from the problems of difficult and slow proliferation *in vitro*; nucleus pulposus cells are relatively easier to culture and proliferate rapidly *in vitro*, but age easily; the anatomical and physiological characteristics of notochordal cell performance is more suitable for the vertebral disc, but there exists difficulty in material selection and the phenotypic proliferation mechanism is unclear; embryonic stem cells at present are not used due to ethical considerations and owing to the failure to solve the problem of how to obtain them from other organs. Also, the differentiation capacity of progenitor cells and the mechanism by which this happens is not very clear; studies are still in the initial stage. A comprehensive comparison of research on MSCs will lead to, relatively speaking, more advantages in terms of its applications. There are some problems that need to be addressed such as cell-specific markers; *in vitro* isolation, culture and proliferation, regulation, and directional differentiation; *in vivo* biological environment adaptability; tumorigenicity; and so on. The co-culturing of progenitor cells and genetic modification of progenitor cells have become the focus of research, which shows that a composite culture or a variety of progenitor cells and their products have better advantages than a single progenitor cell making the seed cell, that is, superior performance is observed, but the project becomes complicated. Therefore, more research is needed to confirm if this is true or if a single seed cell is superior.

# References

1. Clouet J, Vinatier C, Merceron C, *et al.* The intervertebral disc: From pathophysiology to tissue engineering. *Joint Bone Spine*, 2009, 76(6): 614–618.
2. Sheikh H, Zakharian K, De La Torre RP, *et al.* *In vivo* intervertebral disc regeneration using stem cell-derived chondroprogenitors. *J Neurosurg Spine*, 2009, 10(3): 265–272.

3. Feng G, Yang X, Shang H, *et al.* Multipotential differentiation of human annulus fibrosus cells: An *in vitro* study. *J Bone Joint Surg Am*, 2010, 92(3): 675–685.

4. Saraiya M, Nasser R, Zeng Y, *et al.* Reversine enhances generation of progenitor-like cells by dedifferentiation of annulus fibrosus cells. *Tissue Eng Part A*, 2010, 16(4): 1443–1455.

5. Yang X, Wang D, Hao J, *et al.* Enhancement of matrix production and cell proliferation in human annulus cells under bioreactor culture. *Tissue Eng Part A*, 2011, 17(11–12): 1595–1603.

6. Henriksson H, Thornemo M, Karlsson C, *et al.* Identification of cell proliferation zones, progenitor cells and a potential stem cell niche in the intervertebral disc region: A study in four species. *Spine*, 2009, 34(21): 2278–2287.

7. Blanco JF, Graciani IF, Sanchez-Guijo FM, *et al.* Isolation and characterization of mesenchymal stromal cells from human degenerated nucleus pulposus: Comparison with bone marrow mesenchymal stromal cells from the same subjects. *Spine*, 2010, 35(26): 2259–2265.

8. Rutges J, Creemers LB, Dhert W, *et al.* Variations in gene and protein expression in human nucleus pulposus in comparison with annulus fibrosus and cartilage cells: Potential associations with aging and degeneration. *Osteoarthritis Cartilage*, 2010, 18(3): 416–423.

9. Minogue BM, Richardson SM, Zeef LA, *et al.* Transcriptional profiling of bovine intervertebral disc cells: Implications for identification of normal and degenerate human intervertebral disc cell phenotypes. *Arthritis Res Ther*, 2010, 12(1): R22.

10. Hegewald AA, Endres M, Abbushi A, *et al.* Adequacy of herniated disc tissue as a cell source for nucleus pulposus regeneration. *J Neurosurg Spine*, 2011, 14(2): 273–280.

11. Guehring T, Wilde G, Sumner M, *et al.* Notochordal intervertebral disc cells sensitivity to nutrient deprivation. *Arthritis Rheumatism*, 2009, 60(4): 1026–1034.

12. Erwin WM, Las Heras F, Islam D, *et al.* The regenerative capacity of the notochordal cell: Tissue constructs generated *in vitro* under hypoxic conditions. *J Neurosurg Spine*, 2009, 10(6): 513–521.

13. Kim JH, Deasy BM, Seo HY, *et al.* Differentiation of intervertebral notochordal cells through live automated cell imaging system *in vitro*. *Spine*, 2009, 34(23): 2486–2493.

14. Weiler C, Nerlich AG, Schaaf R, *et al.* Immunohistochemical identification of notochordal markers in cells in the aging human lumbar intervertebral disc. *Eur Spine J*, 2010, 19(10): 1761–1770.

15. Quintin A, Schizas C, Scaletta C, *et al.* Isolation and *in vitro* chondrogenic potential of human foetal spine cells. *J Cell Mol Med*, 2009, 13(8B): 2559–2569.

16. Hiyama A, Mochida J, Sakai D. Stem cell applications in intervertebral disc repair. *Cell Mol Biol (Noisy-le-grand)*, 2008, 54(1): 24–32.

17. Sobajima S, Vadala G, Shimer A, *et al.* Feasibility of a stem cell therapy for intervertebral disc degeneration. *Spine*, 2008, 8(6): 888–896.

18. Bieback K, Kinzebach S, Karagianni M. Translating research into clinical scale manufacturing of mesenchymal stromal cells. *Stem Cells Int*, 2011, 2010: 193519.

19. Ryan JM, Barry F, Murphy JM, *et al.* Interferon-gamma does not break, but promotes the immunosuppressive capacity of adult human mesenchymal stem cells. *Clin Exp Immunol*, 2007, 149(2): 353–363.

20. Watanabe T, Sakai D, Yamamoto Y, *et al.* Human nucleus pulposus cells significantly enhanced biological properties in a coculture system with direct cell-to-cell contact with autologous mesenchymal stem cells. *J Orthop Res*, 2010, 28(5): 623–630.

21. Yang SH, Wu CC, Shih TT, *et al. In vitro* study on interaction between human nucleus pulposus cells and mesenchymal stem cells through paracrine stimulation. *Spine*, 2008, 33(18): 1951–1957.

22. Yang H, Wu J, Liu J, *et al.* Transplanted mesenchymal stem cells with pure fibrinous gelatin-transforming growth factor-beta1 decrease rabbit intervertebral disc degeneration. *Spine J*, 2010, 10(9): 802–810.

23. Hohaus C, Ganey TM, Minkus Y, *et al.* Cell transplantation in lumbar spine disc degeneration disease. *Eur Spine J*, 2008, 17(Suppl 4): 492–503.

24. Meisel HJ, Siodla V, Ganey T, *et al.* Clinical experience in cell-based therapeutics: Disc chondrocyte transplantation a treatment for degenerated or damaged intervertebral disc. *Biomol Eng*, 2007, 24(1): 5–21.

25. Momin EN, Vela G, Zaidi HA, *et al.* The oncogenic potential of mesenchymal stem cells in the treatment of cancer: Directions for future research. *Curr Immunol Rev*, 2010, 6(2): 137–148.

26. Herberts CA, Kwa MS, Hermsen HP. Risk factors in the development of stem cell therapy. *J Transl Med*, 2011, 9: 29.

27. Gastaldi G, Asti A, Scaffino MF, *et al.* Human adipose-derived stem cells (hASCs) proliferate and differentiate in osteoblast-like cells on trabecular titanium scaffolds. *J Biomed Mater Res A*, 2010, 94(3): 790–799.

28. Jackson WM, Aragon AB, Djouad F, *et al.* Mesenchymal progenitor cells derived from traumatized human muscle. *J Tissue Eng Regen Med*, 2009, 3(2): 129–138.

29. Murrell W, Sanford E, Anderberg L, *et al.* Olfactory stem cells can be induced to express chondrogenic phenotype in a rat intervertebral disc injury model. *Spine J*, 2009, 9(7): 585–594.

30. Nesti LJ, Jackson WM, Shanti RM, *et al.* Differentiation potential of multipotent progenitor cells derived from war-traumatized muscle tissue. *J Bone Joint Surg Am*, 2008, 90(11): 2390–2398.

31. Tao H, Yu MC, Yang HY, *et al.* Effect of allogenic adipose-derived stem cell transplantation on bone mass in rats with glucocorticoid-induced osteoporosis. *J South Med Univ (Nan Fang Yi Ke Da Xue Xue Bao)*, 2011, 31(5): 817–821.

32. Varshney RR, Zhou R, Hao J, *et al.* Chondrogenesis of synovium-derived mesenchymal stem cells in gene-transferred co-culture system. *Biomaterials*, 2010, 31(26): 6876–6891.

33. Kurose R, Ichinohe S, Tajima G, *et al.* Characterization of human synovial fluid cells of 26 patients with osteoarthritis knee for cartilage repair therapy. *Int J Rheum Dis*, 2010, 13(1): 68–74.

34. Lee SY, Nakagawa T, Reddi AH. Mesenchymal progenitor cells derived from synovium and infrapatellar fat pad as a source for superficial zone cartilage tissue engineering: Analysis of superficial zone protein/lubricin expression. *Tissue Eng Part A*, 2010, 16(1): 317–325.

35. Yoshii T, Sotome S, Torigoe I, *et al.* Isolation of osteogenic progenitor cells from trabecular bone for bone tissue engineering. *Tissue Eng Part A*, 2010, 16(3): 933–942.

36. Goldschlager T, Ghosh P, Zannettino A, *et al.* Cervical motion preservation using mesenchymal progenitor cells and pentosan polysulfate, a novel chondrogenic agent: Preliminary study in an ovine model. *Neurosurg Focus*, 2010, 28(6): E4.

37. Sommar P, Pettersson S, Ness C, *et al.* Engineering three-dimensional cartilage-and bone-like tissues using human dermal fibroblasts and macroporous gelatine microcarriers. *J Plast Reconstr Aesthet Surg*, 2010, 63(6): 1036–1046.

38. Jiang Biao, Li Ming, Cao Ming. Experimental study on heterotopic osteogenesis of rabbit bone marrow mesenchymal stem cells modified by human bone morphogenetic protein 9. *J Sichuan Univ* (Medical Science Edition), 2008, 39(5): 723–727.

39. Jiang J, Fan CY, Zeng BF. Experimental construction of BMP2 and VEGF gene modified tissue engineering bone *in vitro*. *Int J Mol Sci*, 2011, 12(3): 1744–1755.

40. Pan C, Hicks A, Guan X, *et al.* SNL fibroblast feeder layers support derivation and maintenance of human induced pluripotent stem cells. *J Genet Genomics*, 2010, 37(4): 241–248.
41. Galende E, Karakikes I, Edelmann L, *et al.* Amniotic fluid cells are more efficiently reprogrammed to pluripotency than adult cells. *Cell Reprogram*, 2010, 12(2): 117–125.
42. Korecki CL, Taboas JM, Tuan RS, *et al.* Notochordal cell conditioned medium stimulates mesenchymal stem cell differentiation toward a young nucleus pulposus phenotype. *Stem Cell Res Ther*, 2010, 1(2): 18.
43. Purmessur D, Schek RM, Abbott RD, *et al.* Notochordal conditioned media from tissue increases proteoglycan accumulation and promotes a healthy nucleus pulposus phenotype in human mesenchymal stem cells. *Arthritis Res Ther*, 2011, 13(3): R81.

# Chapter 7

# Tissue Engineering and Cell Scaffold

The three main components of tissue engineering include seed cells, signals — signal transducing factors (such as cytokines, enzymes, enzyme inhibitors, and growth factors) and biochemical factors (such as mechanical loading), and scaffold. The role of the cell scaffold is to provide a suitable microenvironment for seed cells to promote cell growth, proliferation, and the synthesis of extracellular matrix. It is one of the three essential components of intervertebral disc tissue engineering (Fig. 1). When designing a scaffold for tissue engineering, many factors need to be considered, for example, the immunogenicity, structural and mechanical properties, biocompatibility and biodegradability, and the method of transplantation [1]. For the selection of scaffold materials suitable for intervertebral disc tissue engineering, the following five characteristics should be considered: conduciveness to growth, proliferation, and synthesis of extracellular matrix of seed cells; good biocompatibility and low immunogenicity; as the normal physiological function of intervertebral disc is to provide mechanical strength, the scaffold must also have plasticity; the scaffold should have a 3D structure similar to human physiological structure, first so that the implanted seed cells can adhere well, second so that it is conducive to nutrient infiltration, and finally

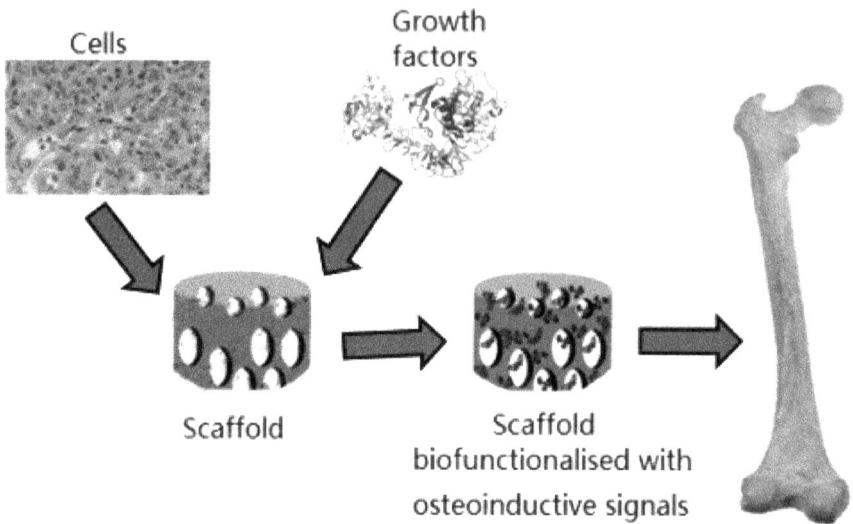

Fig. 1.  The main components of tissue engineering.

to help discharge cell metabolism; and good biodegradability, one of the main conditions that should be fulfilled, since scaffold materials should be gradually decomposed and discharged in the process of tissue formation, so as to minimize the impact of the formation of new tissue.

In intervertebral disc tissue engineering, commonly used carrier materials can be broadly divided into three categories: synthetic materials, natural materials, and composite materials. Natural materials include chitosan, alginate, collagen, fibrin agarose gel; synthesized materials are polymer materials with aliphatic polyester as the main representative, and include polylactic acid, polyglycolic acid copolymer. Composite materials include calcium phosphate, hyaluronic acid. The composite materials also have type I collagen hyaluronic acid, type II collagen hyaluronic acid, polyhydroxy acetic acid hyaluronic acid, malic acid, and so on. In the following, a brief introduction to some of these scaffold materials will be provided.

# 1. Natural Materials

Natural biological materials have been widely used in tissue engineering. The biggest advantage of this type of material is their excellent biological compatibility; at the same time, natural materials have some shortcomings, such as the difficulty of large-scale production, the poor biomechanical performance, immunogenicity, and the risk of virus spread, thus limiting its application. Some representative examples are discussed below.

Alginate: It is a natural linear polysaccharide, is extracted from marine brown algae species, and is negatively charged [2]. When it encounters the two divalent cations (such as $Cu^{2+}$, $Ca^{2+}$) it undergoes an ionic coupling reaction, leading to the formation of a gel material, for example, calcium alginate hydrogel, which has a characteristic structural mesh appearance. Adding a chelating agent (phosphoric acid, EDTA, citric acid and salt, etc.) leads to composite decomposition reaction. Tritz-Schiavi *et al.* and Kim *et al.* [3, 4] postulated that since alginate has low biotoxicity and good biocompatibility, and it is also relatively cheap, it is one of the options for use as tissue engineering carrier material for seed cells to help in intervertebral disc (such as the inner and outer annulus fibrosus cells, nucleus pulposus cells) proliferation and formation of extracellular matrix to provide appropriate local microenvironment. But because the alginate preparation technology is not fully mature yet, alginate is often combined with other materials in experimental studies. Shao and Hunter [5] developed an alginate-chitosan composite scaffold to culture canine intervertebral disc annulus fibrosus cell composite *in vitro*. The SEM results showed that with the prolongation of culture time, the composite scaffold showed slow degradation, with no obvious cytotoxicity and no obvious interference on cell production, and increased growth over a longer period for cells of fibrous ring. It was also found that colonization capacity was good, and the extracellular matrix was abundant.

Chitosan: It is one of the main components in crustacean animal shells. It has biological properties such as biocompatibility, biodegradability, and anti-bacterial activity. The physical and chemical properties of chitosan combined with its special composition makes it a multifunctional biopolymer [6] that is suitable for intervertebral disc tissue engineering. In particular, it is able to form a gel at different pH or temperature. It has been confirmed that chitosan combines with glycerol phosphate or ammonium hydrogen phosphate to produce a kind of heat-induced gel. The gel is a solid at physiological temperature, and it is a liquid at room temperature [7, 8]. The properties of the chitosan gel make it suitable for intradiscal injection. Because the gel has similar mechanical properties to natural tissue, it is recommended as an ideal candidate for nucleus replacement [9]. At present, the existing literature provides evidence for the use of glycerol phosphate/chitosan hydrogel, and literature also shows that bovine nucleus pulposus cell viability and physiological function can be maintained by using such a scaffold [10]. However, different culture media and different culture conditions (microenvironment) can be used to induce human mesenchymal stem cells to differentiate into human nucleus pulposus cells [11].

Sato *et al.* [12] and Gruber *et al.* [13] demonstrated that three kinds of cells in the intervertebral disc (i.e. outer and inner annulus cells and nucleus pulposus cells) were cultured in monolayers and co-cultured using agarose gel composites. They found that altering the glycosaminoglycan content and DNA content led to the formation of one of the three kinds of cells: When glycosaminoglycans were present in monolayer culture, inner annulus cell yield was significantly higher than that of the outer annulus cells (statistically significant); When glycosaminoglycans were present in composite culture, there was no significant difference between the three kinds of cells. Therefore, Sato *et al.* [12] and Gruber *et al.* [13] postulated that for the composite culture of these three kinds of cells, agarose gel is a suitable scaffold material.

Collagen: It is an ideal scaffold material for tissue engineering; it is a kind of natural biological protein that is widely distributed in the animal body, especially in the ligaments, tendons, skin, and so on. It has a characteristic 3 helix peptide structure; this structure has low immunogenicity and excellent biocompatibility, bioactivity, and biodegradability. Gruber *et al.*'s [14] research shows that co-culturing intervertebral disc cells with collagen gel resulted in the growth of cells, circular shape of cells, expression of aggrecan and type I collagen, and presence of type II collagen mRNA; while the intervertebral disc cell morphology at the gel edge was fusiform, there was no expression of type II collagen and proteoglycan mRNA. The experiment indicates that the growth-promoting effect of collagen scaffold on intervertebral disc cells helps to secrete extracellular matrix. Sato *et al.* [15] cultivated rabbit annulus fibrosus cells using collagen as scaffold material in the structure of a honeycomb, and experimental results showed that the growth and proliferation of fibrous ring cells as well as cell phenotype can be maintained; also, co-culturing with scaffold can lead to the production of large amounts of extracellular matrix, and collagen composite fiber ring cell stent implantation can be done for intervertebral disc injury. The experimental results show the formation of hyaline cartilage tissue.

Fibrin gel is a kind of natural polymer formed by thrombin and fibrin polymerization; the polymer has good biocompatibility, biodegradability and absorbability, and low immunogenicity. It can release platelet-derived growth factor and transforming growth factor, so as to promote the adhesion and proliferation of intervertebral disc cells, as well as secrete extracellular matrix, ultimately achieving the function of controlling the growth of cells. Sha'ban *et al.* [16] designed and constructed the PLGA copolymer–fibrin scaffold for the inoculation of annulus fibrosus and nucleus pulposus cells, based on the conventional *in vitro* co-culture techniques, for 3 weeks and observed the results under a light microscope for morphological analysis. The results showed the growth and proliferation of the cells, which were in good condition, the formation of a cartilage-like tissue, and generation of large amounts of extracellular matrix proteins such as collagen, chitosan, and other components. Gruber *et al.* [17] believe that fibrin gel and composite cultured human intervertebral disc cells do not produce aggrecan and 6-chondroitin sulfate sulfotransferase. This enzyme chondroitin sulfate 6 needs to be generated in the process, but the expression of only type I and type II collagen as scaffold material signals obvious defects, so Gruber postulated that fibrin gel scaffold is not suitable for intervertebral disc tissue engineering. Each argument is different, however. The suitability of fibrin gel as a disc tissue engineering scaffold material is still controversial, so it is still necessary for further research to be conducted.

## 2. Synthetic Materials

The advantages of synthetic materials are their better physical and mechanical performance, easy processing, capability of batch production, and their structure and properties which can be controlled, but these kinds of materials do not possess any biological activity, lack cell recognition signals, and can cause aseptic inflammation and other issues. The main representative examples are detailed in the following.

Calcium polyphosphate: calcium dihydrogen phosphate or calcium phosphate is the main raw material, which, through intermolecular polymerization, forms a product that has been widely used in the field of tissue engineering. Calcium phosphate is characterized by its good biological activity, biodegradability, and biomechanical properties, which closely approximate that of the nucleus pulposus. Seguin *et al.* [18] isolated nucleus pulposus cells on the surface of calcium polyphosphate-inoculated materials, cultured them *in vitro* for 6 weeks, and observed them under a light microscope to assess their histology; the collagen and proteoglycan content, DNA content, and biomechanical properties were also measured and analyzed. The results demonstrated that on the surface of calcium polyphosphate, pulp nuclear tissues were detected, and these tissues had structure and biomechanical properties close to the normal nucleus pulposus tissue.

Hyaluronic acid, which is widely distributed in various parts of the body, is not a kind of protein but an acidic polysaccharide with good biocompatibility. Crevensten *et al.* [19] used the hyaluronic acid gel for mesenchymal stem cell culture and transplanted this into a rat for composite intervertebral disc degeneration study. The results showed that 2 weeks after transplantation, the number of mesenchymal stem cells was significantly reduced; interestingly, after 4 weeks of culture, the number of mesenchymal stem cells increased significantly, and also showed better vitality. However, the reason for this change in the number of cells was not clearly explained, and so further research is needed. In the same study, an imaging examination showed that the height of the intervertebral disc also showed an increasing trend. Thus, Crevensten considered that the use of hyaluronic acid gel as a scaffold material is not suitable. In view of the fact that there are few related studies on this topic, it is difficult to determine whether the use of hyaluronic acid as a scaffold material for intervertebral disc tissue engineering is feasible.

Polylactic acid and polyhexyl ester is a kind of synthetic polymer material that has compatibility, excellent plasticity, and good biodegradability. In the process of degradation *in vivo*, polyglycolic acid is generated after the decomposition of lactic acid, both of which can be decomposed into $CO_2$ and water through the respiratory system and urinary system, and so the final product can be discharged. Copolymer of polylactic acid acetic acid and hydroxyl has compatibility, plasticity, and good mechanical strength, is a widely used tissue engineering scaffold material, has better mechanical property, and can be regulated by changing the molecular weight degradation rate. Mizuno *et al.* [20], using polylactic acid and polyglycolic acid as the raw material, seeded purified AF cells on the scaffolds; then after 1D of adherence a solution of sodium alginate (20 g/L) mixed with nucleus pulposus cells was added with $CaSO_4$ and rapidly inoculated in the internal hollow structure of polylactic acid/polyglycolic acid scaffolds and co-cultured *in vitro* following routine procedures. The results are encouraging, showing that 12 weeks after culture, the tissue was formed as assessed with the following aspects: gross morphology, histology, cytology, biochemistry and biomechanics.

## 3. Composite Materials

For the application of composite materials in different combinations, due to the abovementioned characteristics of natural materials and synthetic materials, the preparation requires complex and expensive materials, but this also, to a certain extent, allows for overcoming the inherent shortcomings of a single material. The main representative examples are explained below.

Type I collagen hyaluronic acid: Type I collagen is mainly distributed in the fiber ring tissue, from the outer layer to the inner layer of the fiber ring. The content of type I collagen was significantly increased in the degenerated nucleus pulposus tissue, which was considered as a self-repair reaction, and the content of type I collagen in the normal nucleus pulposus was almost zero. Alini *et al.* [21] used hyaluronic acid and collagen I composite scaffolds for bovine intervertebral disc cell growth on a stent; after 2 months of culturing, conventional *in vitro* experimental data confirmed that in the gel, the cells grew well and secreted extracellular matrix components (such as proteoglycans, type I collagen, etc.).

Type II collagen hyaluronic acid: Huang *et al.* [22] used type II collagen and hyaluronic acid scaffolds after mixing by a chemical covalent binding process using chondroitin sulfate 6 hybrid to grow nucleus pulposus cells for tissue engineering. Rabbit nucleus pulposus cells were implanted onto the scaffold, and histological observation and scanning electron microscopic observation were performed after 4 weeks. The results showed the widespread deposition of extracellular matrix in the mixed scaffold, and the experimental data showed that extracellular matrix proliferation, cell viability, nucleus pulposus proteoglycan content, and activation of related genes showed an increasing trend (genes coding for proteoglycan polymer and type II collagen in the rabbit nucleus pulposus cells were expressed at a higher level); only type I collagen gene expression was low.

Polyhydroxy acetic acid hyaluronic acid: Abbushi and others [23] implanted rabbit intervertebral disc cells by plasma immersion into polyhydroxy acetic acid. The T2 MRI-weighted signal intensity of the experimental group increased by the sixth week after operation to about 45%, while in the control group it only increased by 11%. The results of the histological examination showed cell proliferation in the intervertebral disc defect, that is, promotion of local tissue regeneration, which suggested that the polyglycolic acid–hyaluronic acid scaffold can promote tissue regeneration in intervertebral disc degeneration induced by cartilage formation, but can also promote extracellular matrix protein synthesis, which is favorable for intervertebral disc repair after operation.

Fibrin–hyaluronic acid: Stern [24] studied a fibrin/hyaluronic acid scaffold for the co-culture of nucleus pulposus cells, and we also assessed nucleus pulposus cell alginate scaffold composite culture. Upon a comparative analysis of the two methods, the experimental data show that the fibrin/hyaluronic acid scaffold compound is better than the alginates one as a composite scaffold owing to its better conduciveness, nucleus pulposus cell growth and proliferation, and secretion of extracellular matrix, thus providing the possibility to repair the degeneration of the nucleus pulposus *in vitro*.

Malic acid polyester polymer: Wan *et al.* [25] constructed a malic acid polyester polymer stent for the co-culturing of the stent and fiber rings of rat cells. The results showed that the growth and proliferation of rat annulus fibrosus cells was good, as was the secretion of extracellular matrix (e.g. type II collagen). Recently, Wan *et al.* [26] used malic acid as a scaffold again, and this has been approved by the FDA and is widely used in the tissue engineering research (following the method of polycaprolactone and three alcohol polymerization to obtain a polymer biological material — Poly (Polycaprolactone Triol Malate) (PP-CLM)). PP-CLM has relatively better material biocompatibility and better mechanical strength and plasticity, and the use of bone matrix gelatin and PP-CLM to construct a biphasic scaffold is a unique feature of this stent that can simulate the annulus fibrosus tissue structure. Biomechanical test results show that the tensile strength of the bone matrix gelatin/PP-CLM scaffold is close to the physiological level of the fiber loop of the animal intervertebral disc, which can reach to about 50 times the tensile strength of PP-CLM alone. This is the first report in the literature to study the construction of tissue engineering scaffolds using biphasic biomaterials, and the results are encouraging and inspiring.

Tissue engineering scaffold materials are widely used, and the extracellular matrix is one of the newly emerging ideal materials. The extracellular matrix is mainly secreted by the mesenchymal cells, while stromal cells provide a structure very similar to the normal tissue or organ, with a microenvironment that is highly complex and similar to that seen in normal tissues. Survival, induction, differentiation, and phenotype to maintain the physiological activity of seed cells are important. At the same time, it is imperative to remove the original cell matrix tissue cells, thus eliminating the antigenicity and providing good biocompatibility. The natural extracellular matrix has a complex structure and difficult matrix *in vitro* reconstruction, so the researchers, through continuous testing, mainly using physical or chemical enzymatic methods, tried to obtain acellular extracellular matrix from natural tissue and organs to provide an effective solution for the new problem. Acellular matrix materials have been widely used in the field of tissue engineering and regenerative medicine, including for the induction of stem cell differentiation and the formation of a variety of functional tissues and organs. Because of the special structure of the intervertebral disc, acellular matrix material application to intervertebral disc regeneration and repair had a late start, but now has become a hot research topic. There are different types of extracellular matrix, acellular fibrous ring support, extracellular matrix scaffold, and nucleus pulposus integration into the intervertebral disc cell matrix scaffold.

Scaffold and cytokine: the aim of this combination is to avoid the effect of cytokine release and to overcome the shortcomings of short duration, so that cytokines can play a better role. With the degradation of the stent, it can make the cytokine stable due to the slow release and function in the host cells. At present, it is a hot research topic. Liver phospholipid scaffold and growth factors such as transforming growth factor, bone morphogenetic protein and vascular endothelial growth factor have significant binding ability. Heparin stent can be used to isolate and protect these growth factors, an idea that has merited great attention [27, 28]. There are a number of methods for establishing this [29], which have been published, and the results have been encouraging. In tissue engineering, it is widely used as the functional scaffold for liver phospholipids [30, 31]. Functional scaffolds constructed using liver phospholipids and statins have been reported to promote osteogenic differentiation of human mesenchymal stem cells [30]. In another study, bone morphogenetic protein 2 was bound to the surface of functional liver phospholipids, which greatly promoted the adhesion and differentiation of rat mesenchymal stem cells [31]. Moreover, a chitosan–high iodine oxide–liver phospholipid gel has been manufactured, and it has been shown to stimulate the biological activity of the controlled release of transforming growth factor 2 [31]. Studies have shown that growth factor bone morphogenetic protein 2 and transforming growth factor 2 are beneficial to nucleus pulposus regeneration. Therefore, it is necessary to carry out further research on the liver phospholipid scaffold in tissue engineering.

Intervertebral disc tissue engineering provides a bright future for the treatment of human disc degeneration, but it is still far from clinical application. Most of the *in vivo* experimental data were obtained from rat, rabbit, and canine models, because of the ease of obtaining samples and performing the operations, but in these animal models, the notochord cells have the entire stage for the development potential of nucleus pulposus cells in intervertebral disc, but in humans this occurs only in the embryonic stage of development. Therefore, one should pay attention to the difference between human and animal cells; also, the walking function of these animals is different from that of humans, and so their intervertebral disc is different from humans, even the process of degeneration is not similar. These models can therefore not accurately simulate the process of human intervertebral disc degeneration, thereby precluding the cognition of the mechanism of degeneration. The intervertebral disc tissue engineering is in the initial stage of study and the nucleus pulposus cells and fibrous ring cell growth is too slow.

Other problems include how to obtain the intervertebral disc cells and the difficulties in simulating the biomechanical environment of scaffolds. One of the hotspots and difficulties in the research of intervertebral disc tissue engineering is the research and development of scaffold materials with good performance. With the development of material science, materials can be developed to the nanometer level, with better biological and physicochemical properties; in addition, the principle of bionics in the process of simulating human intervertebral disc tissue on the scaffold material is also a trend that is in development; also, the fact of whether an injectable scaffold can be used is a hot research topic, because it has been injected into the nucleus pulposus during surgical intervention or minimally invasive operations so as to achieve repair. The selection range of injectable scaffold materials mainly include chitosan, type II collagen, hyaluronic acid, protein fiber elastin, and alginate, but chitosan is the scaffold of choice. In brief, the development of intervertebral disc tissue engineering, for the repair of intervertebral disc degeneration, is to restore the physiological function, so as to provide new treatments for degenerative disc diseases, thereby bringing hope and bright prospects.

## References

1. Huang KY, Yan JJ, Hsieh CC, *et al*. The *in vivo* biological effects of intradiscal recombinant human bone morphogenetic protein-2 on the injured intervertebral disc: An animal experiment. *Spine* (*Phila Pa 1976*) 2007, 32: 1174–1180.
2. Taqieddin E, Amiji M. Enzyme immobilization in novel alginate-chitosan core-shell microcapsules. *Biomaterials*, 2004, 25(10): 1937–1945.
3. Tritz-Schiavi J, Charif N, Henrionnet C, *et al*. Original approach for cartilage tissue engineering with mesenchymal stem cells. *Biomed Mater Eng*, 2010, 20(3): 167–174.
4. Kim D, Monaco E, Maki A, *et al*. Morphologic and transcriptomic comparison of adipose- and bone-marrow-derived porcine stem cells cultured in alginate hydrogels. *Cell Tissue Res*, 2010, 341(3): 359–370.

5. Shao X, Hunter CJ. Developing an alginate/chitosan hybrid fiber scaffold for annulus fibrosus cells. *J Biomed Mater Res A*, 2007, 82(3): 701–710.

6. Di Martino A, Sittinger M, Risbud MV. Chitosan: A versatile biopolymer for orthopaedic tissue-engineering. *Biomaterials*, 2005, 26: 5983–5990.

7. Cho J, Heuzey MC, Beìgin A, *et al*. Physical gelation of chitosan in the presence of beta-glycerophosphate: The effect of temperature. *Biomacromolecules,* 2005, 6: 3267–3275.

8. Nair LS, Starnes T, Ko JW, *et al*. Development of injectable thermogelling chitosan-inorganic phosphate solutions for biomedical applications. *Biomacromolecules*, 2007, 8: 3779–3785.

9. Bader RA, Rochefort WE. Rheological characterization of photopolymerized poly (vinyl alcohol) hydrogels for potential use in nucleus pulposus replacement. *J Biomed Mater Res A*, 2008, 86: 494–501.

10. Roughley P, Hoemann C, DesRosiers E, *et al*. The potential of chitosanbased gels containing intervertebral disc cells for nucleus pulposus supplementation. *Biomaterials*, 2006, 27: 388–396.

11. Richardson SM, Hughes N, Hunt JA, *et al*. Human mesenchymal stem cell differentiation to NP-like cells in chitosan-glycerophosphate hydrogels. *Biomaterials*, 2008, 29: 85–93.

12. Sato M, Kikuchi T, Sazuma T, *et al*. Glycosaminoglycan accumulation in primary culture of rabbit intervertebral disc cells. *Spine (Phila Pa 1976)* 2001, 26(24): 2653–2660.

13. Gruber HE, Fisher EC Jr, Desai B, *et al*. Human intervertebral disc cells from the annulus: Three-dimensional culture in agaroseor alginate and responsiveness to TGF-beta1. *Exp Cell Res*, 1997, 235(1): 13–21.

14. Gruber HE, Jolmson T, NDrton HJ, *et al*. The sand rat model for disc degeneration: Radiologic characterization of age-related changes: Cross-sectional and prospective analyses. *Spine (Phila Pa 1976)*, 2002, 27(3): 230–234.

15. Sato M, Kikuchi M, Ishihara M, *et al*. Tissue engineering of the intervertebral disc with cultured annulus fibrosus cells using atelocollagen honeycomb-shaped scaffold with a membrane seal (ACHMS scaffold). *Med Biol Eng Comput*, 2003, 41(3): 365–371.

16. Sha'ban M, Yoon SJ, Ko YK, *et al*. Fibrin promotes proliferation and matrix production of intervertebral disc cells cultured in threedimensional poly (lactic-co-glycolic acid) scaffold. *J Biomater Sci Polym Ed*, 2008, 19(9): 1219–1237.

17. Gruber HE, Leslie K, Ingram J, *et al.* Cell-based tissue engineering for the intervertebral disc: *In vitro* studies of human disc cell gene expression and matrix production within selected cell carriers. *Spine (Phila Pa 1976)*, 2004, 4(1): 44–55.

18. Seguin CA, Grynpas MD, Pilliar RM, *et al.* Tissue engineered nucleus pulposus tissue formed on aporous calcium polyphosphate substrate. *Spine (Phila Pa 1976)*, 2004, 29(12): 1299–1306.

19. Crevensten G, Walsh AJ, Ananthakrishnan D, *et al.* Intervertebal disc cell therapy for regeneration; mesenchymal stem cell implantation in rat intervertebral dics. *Ann Bioned Eng*, 2004, 32(3): 430–434.

20. Mizuno H, Roy AK, Vacanti CA, *et al.* Tissue-engineered composites of annulus fibrosus and nucleus pulposus for intervertebral disc replacement. *Spine (Phila Pa 1976)*, 2004, 29(12): 1290–1297.

21. Alini M, Li W, Markovic P, *et al.* The potential and limitations of a cell-seeded collagen/hyaluronan scaffold to engineer an intervertebral disc-like matrix. *Spine (Phila Pa 1976)*, 2003, 28(5): 446–454.

22. Huang B, Li CQ, Zhou Y, *et al.* Collagen II/hyaluronan/chondroitin-6-sulfate tri-copolymer scaffold for nucleus pulposus tissue engineering. *J Biomed Mater Res B Appl Biomater*, 2010, 92(2): 322–331.

23. Abbushi A, Endres M, Cabraja M, *et al.* Regeneration of intervertebral disc tissue by resorbable cell-free polyglycolic acid-based implants in a rabbit model of disc degeneration. *Spine (Phila Pa 1976)*, 2008, 33(14): 1527–1532.

24. Stern S, Lindenhayn K, Schultz O, *et al.* Cultivation of porcine cells from the nucleus pulposus inafibrin/hyaluronic acid matrix. *Acta Orthop Scand*, 2000, 71(5): 496–502.

25. Wan Y, Feng G, Shen FH, *et al.* Novel biodegradable poly(1,8-octanediol malate) for annulus fibrosus regeneration. *Macromol Biosci*, 2007, 7(11): 1217–1224.

26. Wan Y, Feng G, Shen FH, *et al.* Biphasic scaffold for annulus fibrosus tissue regeneration. *Biomaterials*, 2008, 29(6): 643.

27. Joung YK, Bae JW, Park KD. Controlled release of heparin-binding growth factors using heparin-containing particulate systems for tissue regeneration. *Expert Opin Drug Deliv*, 2008, 5: 1173–1184.

28. Nie H, Wang CH. Fabrication and characterization of PLGA/HAp composite scaffolds for delivery of BMP-2 plasmid DNA. *J Control Release*, 2007, 120: 111–121.

29. Murugesan S, Xie J, Linhardt RJ. Immobilization of heparin: Approaches and applications. *Curr Top Med Chem*, 2008, 8(2): 80–100.

30. Benoit DS, Collins SD, Anseth KS. Multifunctional hydrogels that promote osteogenic hMSC differentiation through stimulation and sequestering of BMP2. *Adv Funct Mater*, 2007, 17(13): 2085–2093.
31. Edlund U, Dånmark S, Albertsson AC. A strategy for the covalent functionalization of resorbable polymers with heparin and osteoinductive growth factor. *Biomacromolecules*, 2008, 9(3): 901–905.

# Chapter 8

# Growth Factors

## 1. Cytokines

Cytokines have an important role in tissue engineering; through endocrine, paracrine, and autocrine function of the system [1], they regulate the metabolic activity of intervertebral disc cells. Studies [2, 3] have shown that there are a series of cytokines that together regulate disc metabolism, and some *in vitro* experiments have also confirmed that these cytokines even play a role in slowing down or reversing the disc degeneration. In the synthesis metabolism, transforming growth factor beta (TGF-β), bone morphogenetic proteins (BMPs), insulin-like growth factor-1 (IGF-1), and Sox-9 were confirmed to promote the synthesis of proteoglycans and type II collagen; in the catabolism reaction, interleukin (IL-1), tumor necrosis factor alpha (TNF-α), and matrix metalloproteinases (MMPs) promote the degradation of proteoglycans and type II collagen.

In 1991, Thompson [4], for the first time, reported that addition of the cytokine TGF-α during *in vitro* culture of nucleus pulposus cells can promote the secretion of protein polysaccharide by these disc cells. Currently, more research is being carried out on BMP family members, including BMP-2 (bone morphogenetic protein-2), -7, and -5, GDF (growth differentiation factor 5), and so on. Kim [5] studying BMP-2 reported predominant chondroitin gene expression in human intervertebral disc cells, increased proteoglycan synthesis, and increased expression and upregulation of type I and II collagen mRNA. BMP-7 is a kind of OP-1 (Osteogenic protein-1) that has been confirmed to have bone inducing activity [6], and it has been found that OP-1 can play an important role in degenerative intervertebral disc repair and can prevent intervertebral disc cell apoptosis. Takegami [7] demonstrated that injection of OP-1 (100 g/per disc) into the nucleus pulposus of degenerated intervertebral discs can enable the promotion of cell proliferation, contribute to the formation of extracellular matrix components such as type II collagen and aggrecan, can promote the expression of *Sox-9* gene, and can also promote the repair of inner annulus. The presence of OP-1 in the nucleus is not confirmed by expression of osteogenic proteins, indicating its role in maintaining the phenotype of chondrocytes in the nucleus pulposus rather than osteogenesis. Research on OP-1 [8] injection into the nucleus pulposus demonstrated inhibition of the expression of MMP-13, TNF-$\alpha$, IL-1 alpha and beta, and reduction in inflammation and pain. OP-1 may also delay disc degeneration, which could be related to the inhibition of aggrecanase. Park *et al.* [9] showed that Fas and FasL can induce the apoptosis of intervertebral disc cells by using the immunohistochemical method. Wei *et al.* [10] found that isolating and culturing nucleus pulposus cells *in vitro*, after pre-processing with OP-1, can significantly prevent apoptosis induction, through blocking the TNF-$\alpha$-induced mechanism and through silencing of Caspase-3. Gruber *et al.* [11] reported that the addition of insulin-like growth factor-1 and platelet-derived growth factor in the nucleus pulposus cells could prevent the apoptosis of cells *in vitro*. GDF-5 is also a member of the MBP family, and the study showed that it can promote the formation of protein polysaccharide and type II collagen, and can improve the proliferation of nucleus pulposus and fiber ring cells [12]. Walsh *et al.*

[13] through the experimental animal model of rat tail intervertebral disc degeneration found that GDF-5 has a main effect on the inner layer of fiber ring cells and can help in nucleus pulposus cell differentiation. Robert *et al.* [14] divided 46 patients requiring surgery according to the classification of degenerative disc disease, as posterior annulus fibrosus tear of intervertebral disc, spondylolisthesis, and lumbar disc herniation. Then, an immunohistochemical method was used to detect the content of matrix metalloproteinases, and it was found that matrix metalloproteinases (MMPs) 1, 2, 3, and 9 are widespread in the lesions in the intervertebral disc. Upon comparison of matrix metalloproteinase levels, the highest level was noted in lumbar disc herniation patients, suggesting that MMPs may play a role in the intervertebral disc degeneration process as a biological regulatory factor involved in the degradation of extracellular matrix of the nucleus pulposus.

In both cell culture experiments *in vitro* and animal experiments *in vivo*, cytokines (playing a major role in the nucleus pulposus cells) showed good effect in repairing the damage, but the majority of experimental animals (rat or rabbit) are young (5–6 months), and hence the influence of notochordal cells on experimental results must also be considered. Therefore, one of the special factors that must be taken into consideration when using experimental animal models is the age of the animal. It is yet to be confirmed whether selecting mature or older (equivalent to 1–4 years) experimental animals (rat or rabbit), in whom the water and polysaccharide content of the nucleus is decreased, would yield similar experimental results.

## 2. Transgenic Continuous Expression of Cytokines

There are some drawbacks to the direct injection of cytokines for degenerative intervertebral disc repair, such as the fact that direct injection can lead to degeneration of nucleus, and also the presence of certain combinations of these cytokines in the cells can sometimes lead to leakage from the cells; in addition to these, the cytokines are proteins that have a set half-life, and so they cannot have long-term effects. How to use the carrier in the degenerated intervertebral disc to enable continued expression of these cytokines has become a hot topic in research [15]. Gene therapy is the process by which the seed cells are transplanted with the target gene and transfected into the degenerated nucleus pulposus, leading to increased expression of specific mRNAs, which, finally, through the paracrine pathway affects the transfected cells and the surrounding cells. The aim is to reduce the imbalance of protein polysaccharides by changing the synthesis of protein polysaccharides and regulating the mode of protein synthesis. The production of type II collagen, as a result of Sox-9, can help achieve the repair of the degenerated intervertebral disc.

At present, the carriers for gene transfection are of two types: viral vectors and non-viral vectors. Viral vectors [16] (retrovirus, adenovirus and adeno-associated virus, and lentivirus), have the potential to cause infection both *in vitro* and *in vivo*, and even non-dividing cells can be infected; also, the virus can infect host cells by integrating into the host DNA, thereby leading to its continuous expression for a long time [17]. The advantage of using a viral vector for transfection includes the low mutation rate, but there are also concerns such as the expression of viral genes, risk of systemic virus infection, and immunogenicity of viral proteins that may stimulate an immune response, thereby affecting the success of the gene therapy. Non-viral vector gene transfer methods include naked DNA injection and the use of plasmids. The transfection efficiency is lower than when using viral vectors, but the advantages include the absence of the side effects associated with virus gene expression and the effects of the immune system (i.e. immune response).

Zhang *et al.* [18] used an adenovirus vector to transfect bone morphogenetic proteins (BMP-2, 3, 4, 5, 7, 8, 10, 11, 12,

13, 14, and 15) and Sox-9 into bovine nucleus pulposus cells and the cells were then cultured *in vitro* for 6D. Nucleus pulposus cells containing proteins BMP-2 and BMP-7 had increased sugar secretion, while those cells with higher BMP-2 had better viability and proliferation. Nucleus pulposus cells containing Sox-9, BMP-4, and BMP-14 showed an increase in collagen secretion, which was more significant compared with the control group (552% and 661%, respectively). Paul *et al.* [19] used a rabbit lumbar disc degeneration model by acupuncture fiber ring method and used a viral vector for injection of Sox-2 into the lesion of nucleus pulposus. Reversal of the degeneration and the restoration of normalcy of the morphology of the nucleus pulposus cells was observed. Wallach [20] successfully transfected TIMP-1 gene using viral vectors into human degenerative nucleus pulposus cells that were clustered in cell culture and were cultured in polypropylene pipe (called "pellet culture" 3D environment culture). Experimental results showed that the transfection of tissue inhibitor of matrix metalloproteinase (TIMP-1), compared to the transfection of BMP-2, into degenerative nucleus pulposus cells led to a significant increase in the level of protein polysaccharide. In the animal model of intervertebral disc degeneration, gene therapy has shown encouraging results, but there are still many problems to be solved [21] such as the construction of a viral vector and the methods to make it more secure, stable, and efficient; other problems include the methods to make gene expression in the host long term and techniques to make its application in humans a reality, so as to achieve the same experimental results seen in animals. To solve all these problems, further research is needed.

## References

1. Boubrlak OA, Watson N, Sivan SS, *et al*. Factors regulating viable cell density in the intervertebral disc: Blood supply in relation to disc height. *J Anat*, 2013, 222(3): 341–348.

2. Masuda K. Biological repair of the degenerated intervertebral disc by the injection of growth factors. *Eur Spine J*, 2008, 17: S441–S451.

3. Masuda K, Imai Y, Okuma M, *et al.* Osteogenic protein-1 injection into a degenerated disc induces the restoration of disc height and structural changes in the rabbit anular puncture model. *Spine (Phila Pa 1976)*, 2006, 31(7): 742–754.

4. Thompson JP, Oegema TJ, Bradford DS. Stimulation of mature canine intervertebral disc by growth factors. *Spine*, 1991, 16(3): 253–260.

5. Kim DJ, Moon SH, Kim H, *et al.* Bone morphogenetic protein-2 facilitates expression of chondrogenic, not osteogenic, phenotype of human intervertebral disc cells. *Spine*, 2003, 28(24): 2679–2684.

6. Hsu WK, Wang JC. The use of bone morphogenetic protein in spine fusion. *Spine J*, 2008, 8(3): 419–425.

7. Takegami K, An HS, Kumano F, *et al.* Osteogenic protein-1 is most effective in stimulating nucleus pulposus and annulus fibrosus cells to repair their matrix after chondroitinase ABC-induced *in vitro* chemonucleolysis. *Spine J*, 2005, 5(3): 231–238.

8. Chubinskaya S, Kawakami M, Rappoport L, *et al.* Anti-catabolic effect of OP-1 in chronically compressed intervertebral discs. *J Orthop Res*, 2007, 25(4): 517–530.

9. Park JB, Kim KW, Han CW, *et al.* Expression of Fas receptor on disc cells in herniated lumbar disc tissue. *Spine*, 2001, 26(2): 142–146.

10. Wei A, Brisby H, Chung SA, *et al.* Bone morphogenetic protein-7 protects human intervertebral disc cells *in vitro* from apoptosis. *Spine*, 2008, 8(3): 466–474.

11. Gruber HE, Norton HJ, Hanley EN Jr. Anti-apoptotic effects of IGF-1 and PDGF on human intervertebral disc cells *in vitro*. *Spine*, 2000, 25(17): 2153–2157.

12. Chujo T, An HS, Akeda K, *et al.* Effects of growth differentiation factor-5 on the intervertebral disc-*in vitro* bovine study and *in vivo* rabbit disc degeneration model study. *Spine*, 2006, 31(25): 2909–2917.

13. Walsh AJL, Bradford DS, Lotz JC. *In vivo* growth factor treatment of degenerated intervertebral discs. *Spine*, 2004, 29(2): 156–163.

14. Roberts S, Caterson B, Menage J, *et al.* Matrix metalloproteinases and aggrecanase: Their role in disorders of the human intervertebral disc. *Spine*, 2000, 25(23): 3005–3013.

15. Nishida K, Suzuki T, Kakutani K, *et al.* Gene therapy approach for disc degeneration and associated spinal disorders. *Eur Spine J*, 2008, 17(supple 4): S459–S466.

16. Hubert MG, Vadala G, Sowa G, *et al.* Gene therapy for the treatment of degenerative disk disease. *J Am Acad Orthop Surg*, 2008, 16(6): 312–319.
17. Miyazaki M, Sugiyama O, Zou J, *et al.* Comparison of lentiviral and adenoviral gene therapy for spinal fusion in rats. *Spine*, 2008, 33(13): 1410–1417.
18. Zhang Y, An HS, Thonar EJ, *et al.* Comparative effects of bone morphogenetic proteins and sox9 overexpression on extracellular matrix metabolism of bovine nucleus pulposus cells. *Spine*, 2006, 31(19): 2173–2179.
19. Paul R, Haydon RC, Cheng HW, *et al.* Potential use of Sox9 gene therapy for intervertebral degenerative disc disease. *Spine*, 2003, 28(8): 755–763.
20. Wallach CJ, Sobajima S, Watanabe Y, *et al.* Gene transfer of the catabolic inhibitor TIMP-1 increases measured proteoglycans in cells from degenerated human intervertebral discs. *Spine*, 2003, 28(20): 2331–2337.
21. Lee JY, Hall R, Pelinkovic D, *et al.* New use of a three-dimensional pellet culture system for human intervertebral disc cells — Initial characterization and potential use for tissue engineering. *Spine*, 2001, 26(21): 2316–2322.

# Chapter 9

# Bone Marrow Mesenchymal Stem Cells and Their Application in Intervertebral Disc Tissue Engineering

With the development of tissue engineering and cell therapy, the key necessity is to find seed cells and target cells that can meet the requirements and are easy to handle. Stem cells, which have high self-renewal capacity and the potential to differentiate into multiple lines, have become the preferred seed cells for tissue engineering. Bone marrow mesenchymal stem cells (BMSCs) are a group of non-hematopoietic stem cells that exist in the bone marrow and can be used as seed cells in tissue engineering owing to their many advantages. First, BMSCs are derived from the bone marrow, which is an adequate source of easy-to-obtain cells; second, BMSCs cultured *in vitro* have the advantages of simple operation, as cell amplification is easy; in addition, the low immunogenicity of BMSCs leads to low levels of allograft rejection; finally, they can also be used as target of cells for gene regulation [1, 2]. These advantages make BMSCs a good candidate in gene therapy, and in recent years

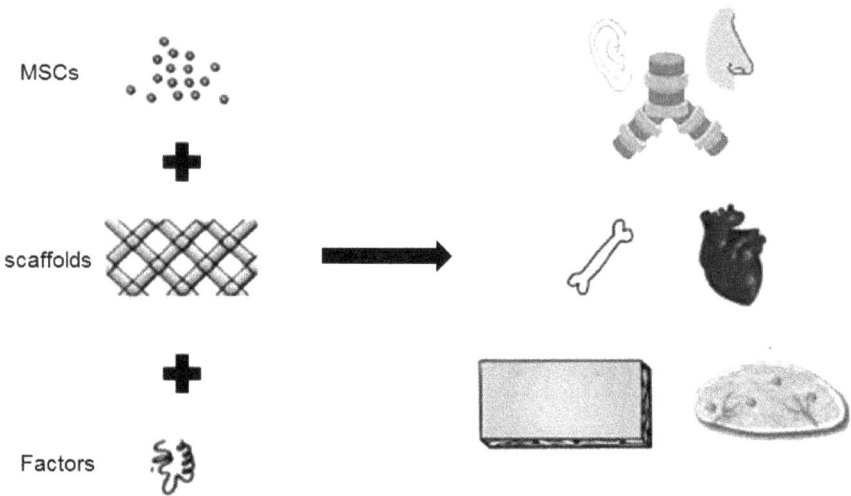

Fig. 1. The multiple lineages of differentiation possible provide a theoretical basis for the widespread application of BMSCs.

this has become a research hotspot in intervertebral disc tissue engineering. This chapter provides a summary of the biological characteristics of BMSCs and also highlights its progress in the application for intervertebral disc tissue engineering (Fig. 1).

## 1. Biological Characteristics of BMSCs

The number of MSCs in the bone marrow is low, for approximately every 105 mononuclear cells there is only 1 MSC, and with increasing age and physical weakness, further reduction in numbers is seen [3]. MSCs have the potential of multi-directional differentiation, spontaneous differentiation, and aging during *in vitro* culture. Studies [4] have shown that, BMSCs can proliferate *in vitro* for about 40 generations, but during passage gradually lose their ability to differentiate into adipocyte and chondrocyte lineages, thus resulting in only osteogenic differentiation. *In vitro* and *in vivo* intervention factors can be used to change the microenvironment to produce various different cells such as bone cells, cartilage cells [5, 6], fat cells [7, 8], myocardial cells, and endothelial cells [9]; mesodermal cell differentiation lineages can also be obtained [10, 11] such as nerve cells, as well as cells of the ectodermal and endodermal lineages, such as liver cells. This differentiation potential provides a theoretical basis for the widespread application of BMSCs. After induction, BMSCs can be differentiated into the corresponding tissues to help repair the defect.

## 1.1. *Cell morphology*

The special conditions of tissue damage can be observed under a light microscope or a phase contrast microscope, wherein BMSCs show a similar fibroblast appearance. When viewed under a transmission electron microscope, BMSCs showed two different morphological structures [12]: (1) relatively quiescent cells with a large, oval nucleus containing only 1 nucleolus, low nuclear cytoplasmic ratio, and presence of few cytoplasmic organelles; and (2) a relatively active group of cells, in which the cell volume is greater than the former type, nucleus is irregular, there is a prominent nuclear membrane with 2–3 small nucleoli, low nuclear cytoplasmic ratio, and abundant organelles in the cytoplasm.

## 1.2. *Cell cycle*

The *in vitro* growth of BMSCs begins with the lag phase, then follows through with the logarithmic growth period, and finally reaches a steady period of growth. Studying the cell cycle, it is evident that only a few BMSCs are in the active replication stage (about 10% in the S + G2 + M phase), and most of the cells are in the stationary phase (G0/G1); the presence of a high proportion of G0/G1-phase cells suggests that BMSCs have high differentiation and subculture potential, high amplification ability, and can maintain the karyotype and telomerase activity of cells [13]. But too much will damage the cultured cells, or even lead to apoptosis of BMSCs [14]. Conget detected BMSCs as making up 20% of the entire number of cells in the G0 phase, which is enough to supply the conditions needed to maintain cell proliferation and differentiation *in vitro*, and even at the 20–25th generation, cell morphology, growth curve, and immunophenotype showed no significant change [14].

### 1.3. *Separation and cultivation of BMSCs*

At present, the main method of separating BMSCs is by density gradient centrifugation and adherence screening method, flow cytometry, and immunomagnetic separation. Flow cytometry and immunomagnetic separation methods have a great impact on the cell activity, and can even lead to complete loss of cell activity. For experimental conditions, large amounts of bone marrow are needed, and so even though the density gradient centrifugation and adherence screening methods are complex, the purity ratio of the adherence screening method is high, and thus density gradient centrifugation method and adherence screening method are now increasingly being used.

The 2D culture model has been widely used, and although it is gradually improving, there are still difficulties to overcome, such as the following [15]: (1) the wall surface area is limited and the cell yield is low; (2) the aseptic operation process is complicated and can easily be contaminated; (3) there is gradual accumulation of metabolites, which can cause cell growth; (4) cells are not in the 3D *in vivo* structure complete with extracellular matrix constituents, thus affecting their biological behavior. Therefore, some researchers put forward the idea of a 3D culture. Glowacki and others [16] cultured cells with collagen sponge as a carrier, which reduced the accumulation of metabolic products. The results showed that the extracellular matrix content was increased and the activity and function of the cultured cells were enhanced. Studies have shown that the shape and quality of these cells [17] are better than those grown in a conventional static culture. The biological reactor can significantly improve the nutritional status of the cultured cells *in vitro* [18]. A biological carrier reactor has also successfully been used in some large-scale animal and plant cell cultures *in vitro*, and the same can be used for BMSC amplification and culture in industrialized settings, so as to meet the demand for the amount of seed cells for cell engineering and gene engineering; but some problems remain to be resolved.

## 1.4. *BMSC identification*

Though the methods for isolating BMSCs are varied, the differentiation potential of different clones is different; even if the purified monoclonal cells are in culture, during culturing the cell morphology and proliferation characteristics of the isolated cells can be modified; and different laboratory conditions lead to different phenotypic expressions of the BMSCs in terms of its differentiation potential. A molecular marker has not yet been found that has strong specificity, and it is generally believed that BMSCs do not express the typical antigens CD34, CD45, and/or HLA-DR that are seen in hematopoietic cells; they also do not express the monocyte/macrophage antigen CD14, the endothelial cell-specific marker CD31, or CD11a, a lymphocyte-specific marker. Antigens include the following categories: (1) adhesion molecules, such as CD166, CD54, CD102, CD44, and CD106; (2) growth factors and cytokines such as interleukin-1 receptor (IL-1R), IL-3R, IL-4R, IL-6R, IL-7R, interferon receptor, and tumor necrosis factor; (3) integrin family members, including CD29, CD49, and CD104; (4) specific antigens, SH2, SH3, SH4, Stro-1, alpha smooth muscle actin, and MAB1740; (5) others, such as CD13, CD71, CD90, CD105. So at present, the identification of BMSCs mainly depends on the exclusion method (assay of antigen labeling) and biological activity analysis (during the course of the culture to provide appropriate stimulation factors to induce differentiation in the mesenchymal stem cells).

## 2. Immunological Properties of BMSCs

### 2.1. *Immunogenicity*

It has been reported [22] that undifferentiated BMSCs have low expression levels of major histocompatibility complex (MHC) I molecules and MHC-II molecules; even if interferon (IFN)-$\gamma$ stimulation can upregulate the expression of MHC-I molecules and MHC-II molecules, BMSCs are seen as allogeneic cells after stimulation of lymphocytes, and thus lymphocytes still cannot produce an immune response. BMSCs thus show low immunogenicity and do not induce acute rejection.

### 2.2. *Immune regulation*

BMSCs can inhibit the cell proliferation induced by mitogens *in vitro*; lymphocytes *in vivo* can inhibit the proliferation and induce immune tolerance [19] and alter the survival time [20]. Also, the time at which BMSCs are collected from the donor can cause acute graft-versus-host disease and prolong the time taken to reach the recipient, suggesting that BMSCs have immunosuppressive effects *in vitro* and *in vivo*. Experiments have reported that [21, 22] this kind of immune suppression is neither immunity nor immune tolerance, but the action of BMSCs on the inhibition of T cells. Therefore, BMSCs have the function of immune regulation and can be used in allograft transplantation, and it can also be used as a safe and effective gene therapy vector.

## 3. Application of BMSCs in Tissue Engineering of Intervertebral Discs

BMSCs, both in *in vitro* and *in vivo* settings, under appropriate conditions of the new microenvironment can be transformed into cartilage cells or nucleus pulposus cells; due to differences in the characteristics, changes in the microenvironment may lead the BMSCs to differentiate into different lineages such as intervertebral disc cells. Sakai and others [23], for the first time, used BMSCs as the seed cells for intervertebral disc tissue engineering. With Ad-lacZ-labeled rabbit autologous BMSCs, a collagen gel composite carrier was injected into the intervertebral space in the region of degeneration; within 4 weeks, staining of the tissue sections confirmed that the transplanted BMSCs were still vibrant, and it was also observed that the cell transplantation model enabled the differentiation from BMSCs to become cells similar to that found in the normal intervertebral disc, i.e. spindle cells that are longitudinal, with lamellar arrangement of cells. The study also demonstrated that there was a correlation between the total protein content and the matrix.

The transplanted cells significantly regenerated the intervertebral disc, and the intervertebral disc formed maintained the normal structure and also had the added ability to delay degeneration to the maximum [24]. Crevensten *et al.* [25] used 15% siloxane gel as a carrier for the fluorescently labeled BMSCs and injected them into the seventh caudal intervertebral disc of the rat. Fourteen days after injection, stem cells still existed in the injection disc area, but their number significantly decreased by 28 days; when the number of stem cells was restored to the original number, the survival rate was 100% compared with the control group. The injection rate was highly increased in the experimental group, which thus showed an increase in disc and matrix synthesis; thus, the results showed that BMSCs in the rat intervertebral disc survived. Risbud *et al.* [26] studied rats in which BMSCs were inoculated into the alginate gel and injected into animals; the cells had been co-cultured with TGF-beta 1 under either normal or hypoxic conditions; it was found that hypoxia and TGF-beta 1 stimulate high expression of metalloproteinase 2 and type II collagen, and also promoted the stem cells to differentiate into nucleus pulposus.

Yamamoto cocultured BMSCs and nucleus pulposus cells, and the results showed that this promoted the proliferation of nucleus pulposus cells and matrix synthesis [27]. Sobajima *et al.* [28] co-cultured BMSCs with nucleus pulposus cells to detect the feasibility of BMSCs for the treatment of intervertebral disc degeneration; the conclusion was that BMSCs have good proliferation ability in the intervertebral disc and that they could successfully carry therapeutic genes into the intervertebral disc. The Le Visage *et al.* [29] study confirmed that a co-culture of BMSCs and fiber ring cells led to the expression of a high level of glucose, and also the co-culture of BMSCs and fiber ring cells can improve the synthesis of protein.

Sakai *et a1.* [30] found that 24 weeks after transplantation of BMSCs, the transplantation group animals' degenerative intervertebral disc height was about 91% and the signal intensity on MRI was about 81%, whereas in the sham operation group, the degenerative disc height was about 67% and the signal intensity on MRI was about 60%. This study showed that the intervertebral disc transplantation of BMSCs retained the ring structure of the nucleus, whereas in the sham operation group the intervertebral disc nucleus pulposus was indistinguishable; immunohistochemistry and gene expression analysis indicated that transplantation of BMSCs into the discs led to the proteoglycan accumulation recovery.

In China, Zhao Ziru *et al.* [31] used BMSCs with TGF-α and sodium hyaluronate for implantation in a rabbit model of intervertebral disc degeneration; then the content of polysaccharide proteins was assessed at 2, 4, 6, and 8 weeks with benzene three phenol spectrophotometry, and an immune determination method was used to identify type II collagen content change. The primary culture and subculture of the rabbit bone marrow mesenchymal stem cells demonstrated active proliferative ability, and it was also observed that the content of proteoglycan and collagen type II in the experimental group after 8 weeks was significantly higher than in the model group, thus delaying the degeneration of the intervertebral disc.

## 4. The Use of BMSCs in Intervertebral Disc Tissue Engineering: Problems and Prospects for Application

At present, the experimentation on and clinical application of BMSCs have made great progress, but the research in this field is still in the exploratory stage. The following problems still exist: (1) BMSCs isolated and purified from *in vitro* experiments showed that BMSCs were mixed with various cells, and so there is a problem of how to effectively obtain pure BMSCs, by assessing the cytological characteristics and various stages of differentiation of BMSCs using marker compounds for further study. (2) The process of BMSCs proliferation and differentiation requires appropriate conditions, so as to control the proliferation and avoid the formation of tumors [32, 33], and to start the desired differentiation path at the appropriate time. This needs to be further studied. (3) Improvement of the conversion rate of cells is yet another problem. (4) The current research is limited to studies done on young animals, and so a study of clinical application of BMSCs needs to be done in older animals with advanced clinical stage. (5) Most of the researches cannot explain the long-term effects of donor BMSCs in the recipient, and so its safety is still a concern, but one of the advantages of BMSCs is that it is effective as a carrier for cell therapy and gene therapy purposes for certain diseases.

In short, with the continuous deepening of the study of BMSCs and their application in intervertebral disc tissue engineering, the results are encouraging. But at the same time, there are many specific problems that need to be solved. In the near future, with further research on BMSCs, they can be used as an important source of cells for treating intervertebral disc degeneration, which serves as a new clinical solution for intervertebral disc degeneration after discectomy.

# References

1. Deans RJ, Moseley AB. Mesenchymal stem cells: Biology and potential clinical uses. *Exp Hematol*, 2000, 28(8): 875–878.
2. Yoo JU, Mandell I, Angele P, *et al*. Chondlogenitor cells and gene therapy. *Clin Orthop*, 2000, 379(Suppl): 164–170.
3. David CC, Reiner C, Catla M, *et al*. Rapid expansion of recycling stems cells in cultures of plastic-adherent cells from human bone marrow. *Proc Natl Acad Sci USA*, 2000, 97(7): 3213–3218.
4. Lu Yanmeng, Fu Wenyu, *et al*. Ultrastructure of mesenchymal stem cells derived from human bone marrow. *Chin J Electron Microscopy*, 2002, 21 (): 373–.
5. Muraglia A, Cancedda R, Quarto R, *et al*. Clonal mesenchymal progenitors from human bone marrow differentiate *in vitro* according to a hierarchical model. *J Cell Sci*, 2000, 113(7): 1161–166.
6. Heino TJ, Hentunen TA, Vaananen HK. Conditioned medium from osteocytes stimulates the proliferation of bone marrow mesenchymal stem cells and their differentiation into osteoblasts. *Exp Cell Res*, 2004, 294(2): 468–468.
7. Kim KW, Lim TH, Kim JG, *et al*. The origin of chondrocytes in the nucleus pulposus and histologic findings associated with the transition of a notochordal nucleus pulposus to a fibrocartilaginous nucleus pulposus in intact rabbit intervertebral discs. *Spine*, 2003, 28(10): 982–990.

8. Janderova I, McNeil M, Murrell AN, *et al.* Human mesenehymal stem cells as an *in vitro* model for human adipogenesis. *Obes Res*, 2003, 11(1): 65–74.

9. Itescu S, Schuster MD, Kocher AA. New directions in strategies using cell therapy for heart disease. *J Mol Med*, 2003, 81(5): 288–296.

10. Gu Lili, Peng Guizu, Liu Dewu. Research progress of bone marrow mesenchymal stem. *Foreign Med Sci Biomed Eng — Cells into Liver Cells*, 2005, 28(2): 113–116.

11. Lee KD, Kuo TK, Whang-Peng J, *et al. In vitro* hepatic differentiation of human mesenchymal stem cells. *Hepatology*, 2004, 40(6): 1275–1284.

12. Dezawa M, Kanno H, Hoshino M, *et al.* Special induction of neuronal cells from bone marrow stromal cells and application for autologous transplantation. *J Clin Invest*, 2004, 113(12): 1701–1710.

13. Tamir A, Petrocelli T, Stetler K, *et al.* Stem cell factor inhibits erythroid differentiation by modulating the activity of G1-cyclin-dependent kinase complexes: A role for P27 in erythroid differentiation coupled G1 arrest. *Cell Growth Differ*, 2000, 11(3): 269–277.

14. Conget PA, Minguell J. Phenotypical and functional properties of human bone marrow mesenchymal progenitor cells. *J Cell Physiol*, 1999, 181(1): 67–73.

15. Aung T, Miyoshi H, Tun T, *et al.* Chondroinduction of mouse mesenchymal stem cells in three-dimensional highly porous matrix scaffold. *J Biomed Mater Res*, 2002, 61(1): 75–82.

16. Glowacki J, Mizuno S, Greenberger JS. Perfusion enchances functions of bone marrow stromal cells in three-dimensional culture. *Cell Transplant*,1998, 7(3): 319–326.

17. Solchaga LA, Seidel J, Zeng L, *et al.* Bioreactors mediate the effectiveness of tissue engineering scaffolds. *J Faseb*, 2002, 16(12): 1691–l694.

18. Sikavitsas VI, Bancroft GN, Mikos AG. Formation of three-dimensional cell/polymer constructs for bone tissue engineering in a spinner flask and a rotating wall vessel bioreactor. *J Biomed Mater Res*, 2002, 62(1): 136–148.

19. Zhao RC, Liao L, Han Q. Mechanisms of and perspectives on the mesenchymal stem cell in immunotherapy. *J Lab Clin Med*, 2004, 143(5): 284–291.

20. Li Haowei, Wen Gummei, Xiao Qingzhong, *et al.* Effects of cotransplantation of donor-derived bone marrow mesenchymal stem cells on acute graft versus host disease. *China Pathophysiol*, 2003, 19(5): 577–580.

21. Tse WT, Pendleton JD, Beyer WM, *et al*. Suppression of allogeneic T-cell proliferation by human marrow stromal cells: Implications in transplantation. *Transplantation*, 2003, 75(3): 389–397.

22. Le Blanc K, Tanamik C, Rosendahl K, *et al*. HLA expression and immunologic properties of differentiated and undifferentiated mesenchymal stern ceils. *Exp Hematol*, 2003, 31(10): 890–896.

23. Sakai D, Mochida J, Yamamoto Y, *et al*. Transplantation of mesenchymal stem cells embedded in atelocollagen gel to the intervertebral disc: A potential therapeutic model for disc degeneration. *Biomaterials*, 2003, 24(20): 3531–3541.

24. Sakai D, Mochida J, Iwashina T, *et al*. Differentiation of mesenchymal stem cells transplanted to a rabbit degenerative disc model: Potential and limitations for stem cell therapy in disc regeneration. *Spine J*, 2005, 30(21): 2379–2387.

25. Crevensten G, Walsh AJ, Ananthakrishnan D, *et al*. Intervertebral disc cell therapy for regeneration: Mesenchymal stem cell implantation in rat intervertebral discs. *Ann Biomed Eng*, 2004, 32(3): 430–434.

26. Risbud MV, AIbert TJ, Guttapalli A, *et al*. Differentiation of mesenchymal stem cells towards a nucleus pulposus — like phenotype *in vitro*: implications for cell-based transplantation therapy. *Spine J*, 2004, 29(23): 2627–2632.

27. Yamamoto Y, Mochida J, Sakai D, *et al*. Reinsertion of nucleus pulposus cells activated by mesenchymal stem cells using coculture method decelerated intervertebral disc degeneration. *Spine J*, 2004, 29(14): 1508–1522.

28. Sobajima S, Shimer A, Kim J, *et al*. Feasibility of stem cell therapy for intervertebral disc degeneration. *Spine J*, 2004, 4(2): 117–125.

29. Le Visage C, Kim SW, Tateno K, *et al*. Interaction of human mesenchymal stem cells with disc cells: Changes in extracellular matrix biosynthesis. *Spine J*, 2006, 31(18): 2036–2042.

30. Sakai D, Mochida J, Iwashina T, *et al*. Regenerative effects of transplanting mesenchymal stem cells embedded in atelocollagen to the degenerated intervertebral disc. *Biomaterials*, 2006, 27(3): 335–345.

31. Zhao Ziru, Wu Xiaotao, Qi Yabin. TGF-beta 1 intervention in the body of rabbit bone marrow mesenchymal stem cells on intervertebral disc degeneration treatment experimental study. *Chin Orthopedic J*, 2006, 14(13): 1019–1022.

32. Serakinci N, Guldberg P, Burns JS, *et al.* Adult human mesenchymal stem cell as a target for neoplastic transformation. *Oncogene*, 2004, 23(29): 5095–5098.

33. Ruble D, Garcia-Castro J, Martin MC, *et al.* Spontaneous human adult stem cell transformation. *Cancer Res*, 2005, 65(8): 3035–3039.

# Chapter 10

# Silk Fibroin in the Tissue Engineering of Intervertebral Disc Annulus

Low back pain (LBP) is a common disease seen by physicians in the department of orthopedics. In Asian countries, especially in China, low back pain is one of the main reasons leading to the loss of labor productivity, because these countries have a higher incidence of laborers [1]. Intervertebral disc degeneration (IDD) is one clinical cause of back pain, and is the leading cause of the loss of working ability. IDD is the degeneration of the intervertebral disc, which is a chronic pathological process that not only affects the performance of the disc and its morphology but also changes the properties and molecular composition of intervertebral disc, including changes in nucleus pulposus cells; this process is irreversible [1–4].

Existing clinical treatments have failed to achieve a truly functional restoration and reconstruction of degenerative intervertebral discs [5]. Surgical treatments by removing the lesions of the intervertebral disc tissue or by sacrificing the mobility of spinal segments to relieve the symptoms do not fundamentally solve the biological aspects of intervertebral disc degeneration caused by the loss. In recent years, the use of scaf-folds for tissue engineering to restore intervertebral disc structure and function helps reverse the pathological changes at the cellu-lar level of repair and reconstruction in order to achieve the normal structure and function of the intervertebral disc, and so this could be an ideal treatment method for degenerative interver-tebral discs [3, 6, 7].

Tissue engineering includes three elements: specific tissue cells (seed cells), scaffold materials, and growth factors. Scaffolds play a central role, as they not only provide structural support for particular cells but also serve as a template, guiding tissue regen-eration and helping to control the structure [8]. Silk fibroin is a natural material that has the characteristics of stable structure, high mechanical strength, good biocompatibility, and controllable biodegradation rate. Thus, its application prospect in tissue engi-neering is promising [9–11].

# 1. Fiber Ring Tissue Engineering

## 1.1. *Characteristics of tissue structure of intervertebral disc annulus*

The fiber ring (annulus fibrosus) is composed of collagen fibers and fibrous cartilage and is divided into three layers. The outer layer is mainly made up of fibroblasts: type I collagen is expressed, and its main function is to maintain the fiber ten-sion. In the inner cartilage, type II collagen is expressed and its main function is to bear the pressure of the intervertebral disc. Fibers are arranged in concentric rings in the vertebral body obliquely, and the collagen fibers were arranged in oblique lumbar orientation, with two adjacent layers between the fib-ers forming an angle of approximately 60 degrees [12, 13]. The formation of cross-links leads to a special structure being formed, which has strong anti-tensile ability, can prevent nucleus pulposus protrusion, and, at the same time, allow for spinal motion in the nucleus during motion and sports activi-ties (i.e. control spinal movement).

The front part of the fiber ring is wider than the rear. The gap between the plates is larger, the rear part of the fiber ring is relatively thin, and the plate is dense. The plate can easily break under violent conditions or degeneration. The self-repair ability of degenerative intervertebral disc is very limited, and the main reason for this is the requirement of fiber ring tissue tension, which is based on the mechanical properties of the fiber ring, and the characteristic need of nucleus pulposus to maintain suf-ficient elasticity, and also the deformation can be transferred to the load axial fiber ring. The nucleus degenerates and annulus ruptures, leading to a break in the nucleus lateral sealing struc-ture; the intervertebral disc cannot bear the external force that is applied, and so the ability to adapt to the external changes is greatly reduced. The poor blood supply in the nucleus pulposus, thus leading to a lack of support nutrients, will gradually lead to deterioration accompanied by degeneration. In this case, the repair of the intervertebral disc can only be done by external intervention. Tissue engineering, as a new discipline, has out-standing advantages for the regeneration of tissues and organs and provides a broad platform for the repair and regeneration of intervertebral discs.

## 1.2. *The scaffold materials for tissue engineering of intervertebral disc annulus*

For use as a 3D scaffold in tissue engineering, the support material must have the following characteristics: (1) good biocompatibility and appropriate degradation rate, and the rate of degradation must match with the tissue regeneration rate; (2) dimensional connectivity and good pore structure, with porosity being higher than 90%; (3) good plasticity to facilitate the release of the carrier's active sub-stances; (4) good material–cell interface property that supports cell adhesion, growth, proliferation, and differentiation. Therefore, the material selected for forming a fiber ring support must not only meet the abovementioned criteria but also be similar with regard to the composition, shape, structure, and mechanical properties of the fiber ring cell matrix [6, 8, 10, 12].

### 1.2.1. *Materials currently used for tissue engineering of the fiber ring*

There are three categories of materials that can be used: natural materials, synthetic materials, and composite materials. The natural materials commonly used include demineralized acellular bone matrix [15], collagen [16], alginate, and fibrin gel. Synthetic materials include calcium phosphate, hyaluronic acid, polyamide [17], and polycaprolactone [18]; composite material include Polylactic acid/beta tricalcium phosphate/hyaluronic acid [19], and polycaprolactone/hydrogel. There is a recent rise in the concept of integrated construction, that is, the integration of intervertebral disc scaffold using different materials, such as acellular demineralized bone matrix/acellular matrix of intervertebral disc nucleus [20], thus providing a support that is closer to the natural structure. However, in fiber ring structures made of these materials, there is a lack of biocompatibility, poor mechanical properties, and so on, and so this still needs to be optimized.

## 2. Silk Fibroin as the Repair Material for Intervertebral Disc Annulus

### 2.1. *The physical and chemical properties of silk fibroin protein derived from a variety of insect cocoons*

The silk fibroin can be extracted from a silkworm cocoon. The silk is composed of 20–30% sericin fibroin and 70–80% trace pigments and carbohydrate. Silk fibroin aqueous solution [9] is formed using the silk fibroin by dialysis. Silk fibroin fiber is a kind of natural fiber with high molecular weight and physiological activity. It contains a heavy chain (H-chain, 350 kDa) and a light chain (L-chain, 25 kDa), and an auxiliary protein (30 kDa) [21]. It has 18 basic amino acids, of which glycine, alanine, and serine account for more than 80%. The silk fibroin p-sheet has a more stable structure, is insoluble in water, has excellent mechanical properties, meets the requirements [22] for mechanical strength of the scaffold, and, by simple chemical modifications, its performance can be improved [11, 23, 24]. A tissue engineering scaffold material also must have high porosity and connectivity. The silk fibroin can be used as scaffold material with paraffin microspheres and NaCl particles as porogen [9] by simply processing after freeze drying to form scaffolds and controlling for pore size, connectivity, and support while still maintaining good mechanical strength (Fig. 1).

The silk fibroin cell attachment rate and proliferation rate were good, and the material did not affect the activity of cells; cells could maintain their normal form and function, the biocompatibility of collagen and the effect is good [25, 26], but its mechanical performance is much better than that of collagen. Silk fibroin also has good biological degradation, which is consistent with the growth of the fiber ring cells. Research shows that the degradation rate of scaffolds and the beta folding structure [22] lead to general degradation of silk fibroin materials in 6 months; the degradation rate can be adjusted by changing the ratio of the p-sheet. The decomposition products are non-toxic and can be absorbed by the body.

## 2.2. *Silk fibroin as a fiber ring scaffold*

Silk fibroin as scaffold for tissue engineering can well support fiber ring cell adhesion, growth, and reproduction, once the correct seed cells are selected [17, 27]; fiber ring cells can form

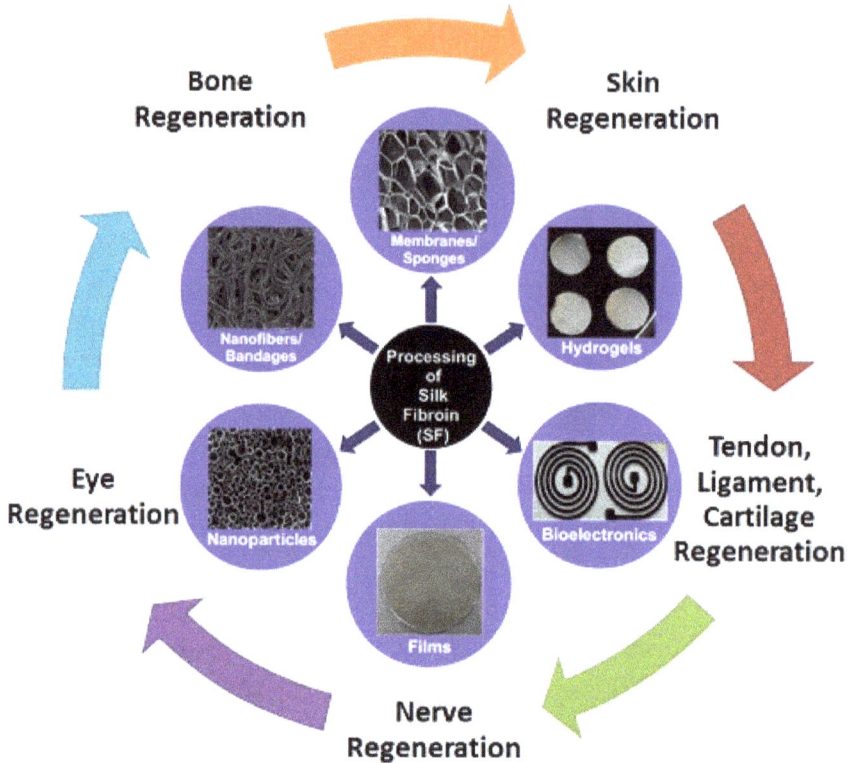

Fig. 1. Silk fibroin, a natural material, has the characteristics of stable structure, high mechanical strength, good biocompatibility, and controllable biodegradation rate. Thus, it has a very good application prospect in tissue engineering. Reprinted with the Creative Commons Attribution License (CC BY 4.0) from Ref. 35.

various stem cells [28, 29], which are found in adequate quantity, are simple to operate, and have numerous application prospects. Chang *et al.* [11] seeded the isolated oxtail intervertebral disc cells on an RGD peptide (arginine–glycine–aspartic acid peptide)-modified silk fibroin scaffold, cultured them for eight weeks, and found that the growth of annulus fibrosus cells in porous silk fibroin scaffold was high in cells with high expression of type II collagen and chitosan. The study also found that the coupling of silk fibroin scaffold with RGD peptide can influence the phenotype of annulus fibrosus cells and that it is favorable for the differentiation of intervertebral disc outer ring.

Research shows that silk fibroin plasticity is good, can guarantee mechanical properties with high porosity and connectivity, to facilitate the growth of cells, and is more conducive to the fiber ring tissue remodeling. Chang *et al.* [10] used sodium chloride particles of different sizes, made corresponding apertures on the silk fibroin scaffold (200 μm, 600 μm, 1000 μm), and then seeded annulus fibrosus cells in the scaffold. Then the structures were divided into static and dynamic training groups, and training was done in different ways: the static culture group was implanted with 20 cells, with a cell density of $2.5 \times 107/mL$, a total of 4 times, every half hour; the dynamic culture group was fixed based on the rotation of the flask, and implanted with the same number of cells, and during the whole training process the scaffolds and cells were maintained in the rotating state, at three different speeds, 60, 90, and 120 r/min. The two experimental groups were cultured to 5 days, adjusting the serum concentration and adding vitamin C. After 2 weeks of training, the amount of collagen and proteoglycan and DNA were much higher in the dynamic group than that of the static group, and this was especially obvious in the 90 r/min cells; the expression of type I collagen in the scaffold was more in cases where the aperture was about 600 μm ECM. The results showed that the silk fibroin scaffold could promote cell quantity, protein polysaccharide amount, and collagen production in the annulus fibrosus tissue and also enhance the structure of the annulus fibrosus tissue and accelerate the remodeling of the fiber ring.

Silk fibroin has good compatibility with other materials, which has important significance for the development of a variety of composite scaffolds. A disc holder made of silk fibroin and silicone resin to ease the integration was made by See *et al.* [30]; then the extracted bone marrow mesenchymal stem cells were seeded in the scaffold, and the cells were cultured under suitable conditions for 4 weeks. Histological observation showed that BMSCs adhered well on the silk fibroin scaffolds and extracellular matrix (ECM) accumulation increased as did the protein content of polysaccharides. And in the training process, the proportion of type II collagen in the BMSC layer in ECM appeared to obviously increase, showing that the provision of a support for annulus fibrosus repair, especially the repair of annulus fibrosus loop, is very effective for reshaping the overall function of the intervertebral disc and is very important.

Park *et al.* [31] made silk fibroin scaffolds in two different groups with a layered and porous morphology (the structure has an outer diameter of 8 mm, an inner diameter of 3.5 mm, and a height of 3 mm (in the layered scaffold, the interlayer distance is 150–250 μm and the average pore size of porous scaffolds is 100–250 μm)) with fibrin/hyaluronic acid (HA) gel for the nucleus pulposus tissue. The scaffolds and cells were cultured for 6 weeks. Histological observation showed that both kinds of structures demonstrated cell adhesion and proliferation; chemical analysis showed that the content of DNA, glycosaminoglycan, and collagen in the lamellar scaffold group was much higher than that in the porous scaffold group; PCR showed that the induced differentiation of COLIA1 fiber ring and aggrecan gene expression were also significantly higher than that of the porous scaffolds group; the mechanical property test showed that the mechanical strength of both the scaffolds was high. These results showed that silk fibroin as a scaffold material is beneficial to the regeneration of the fiber ring tissue, and the lamellar scaffold was much closer in terms of properties to the normal fiber ring, and also led to better repair.

Bhardwaj and others [33] modeled the structural characteristics of the fiber loop, and the silk fibroin fibers were processed to create a structure similar to the fiber board, and this was then used to support the growth of human cartilage cells. The experimental results showed that the chondrocytes grew well, there was a large amount of cartilage matrix deposition, and the fibers arranged in one direction, which is very similar to the structure of a normal fiber ring. The success of this growth of cartilage cells also provides a lot of data to support the fiber ring tissue engineering. In addition, when chondroitin sulfate was added, a cross-linking was observed in the structure, greatly increasing the structural strength, which is very similar to the mechanical properties of the normal intervertebral disc.

Silk fibroin is currently a very interesting material in tissue engineering, with an adequately large source, simple extraction process, stable structure, biological compatibility, and controllable degradation and also has the added characteristics of superior mechanical properties, long life, good plasticity, and the capability to have enhanced performance by a simple chemical modification, all of which are in line with the requirements for a scaffold material to be used for fibrous ring tissue engineering. Thus, silk fibroin is an ideal tissue engineering scaffold material. The combination of silk fibroin with different kinds of growth factors will be more beneficial to the growth of cells and the recovery of tissue function. In addition, pure silk fibroin materials are not able to meet the overall requirements of intervertebral disc tissue engineering, and so integrating other materials will lead to wider application prospects [33, 34]. At present, the research on silk fibroin scaffold has just started, and many problems need to be solved. The frame structure of silk fibroin scaffold needs to be further optimized, because creating animal models of intervertebral disc defect is very difficult. The human intervertebral disc has a very complex micro structure, and must function perfectly; therefore, to fully simulate the same in an *in vivo* environment is not easy, as it requires the combination of various fields such as biology, engineering, material science, basic medicine, clinical medicine, and other disciplines, to help gradually solve the problems in tissue engineering. In short, silk fibroin material is still in the initial stage of development, but it has shown good prospects for the development of intervertebral disc annulus tissue engineering, and so is expected to be widely used in the near future.

# References

1. Whatley BR, Wen HJ. Intervertebral disc (IVD): Structure, degeneration, repair and regeneration. *Mat Sci Eng*, 2012, 2: 61–77.
2. Zhuang Y, Huang B, Li CQ, *et al.* Construction of tissue-engineered composite intervertebral disc and preliminary morphological and biochemical evaluation. *Biochem Biophys Res Commun*, 2011, 2: 327–332.
3. Bron JL, Helder MN, Meisel HJ, *et al.* Repair, regenerative and supportive therapies of the annulus fibrosus: Achievements and challenges. *Eur Spine J*, 2008, 3: 301–313.
4. Schollum ML, Robertson PA, Broom ND. How age influences unravelling morphology of annular lamellae — a study of interfibre cohesivity in the lumbar disc. *J Anatomy*, 2010, 3: 310–319.
5. Meisel HJ, Siodla V, Ganey T, *et al.* Clinical experience in cell-based therapeutics: Disc chondrocyte transplantation A treatment for degenerated or damaged intervertebral disc. *Biomol Eng*, 2007, 1: 5–21.
6. Jin L, Shimmer AL, Li X. The challenge and advancement of annulus fibrosus tissue engineering. *Eur Spine J*, 2013: 1– 11.
7. Endres M, Abbushi A, Thomale UW, *et al.* Intervertebral disc regeneration after implantation of a cell-free bioresorbable implant in a rabbit disc degeneration model. *Biomaterials*, 2010, 22: 5836–5841.
8. Qin Jianghui, Wang Junfei, *et al.* History of winter and spring, with silk scaffold materials research progress. *China Ligament Orthopedic Journal*, 2008, 10: 763–764, 783.
9. Mandal BB, Park SH, Gil ES, *et al.* Multilayered silk scaffolds for meniscus tissue engineering. *Biomaterials*, 2011, 2: 639–651.

10. Chang G, Kim HJ, Vunjak-Novakovic G, *et al.* Enhancing annulus fibrosus tissue formation in porous silk scaffolds. *Biomed Mater Res A*, 2010, 1: 43–51.

11. Chang G, Kim HJ, Kaplan D, *et al.* Porous silk scaffolds can be used for tissue engineering. *Eur Spine J*, 2007, 11: 1848–1857.

12. Nerurkar NL, Mauck RL, Elliott DM. ISSLS Prize winner: Integrating theoretical and experimental methods for functional tissue engineering of the annulus fibrosus. *Spine*, 2008, 25: 2691–2701.

13. Driscoll TP, Nakasone RH, Szczesny SE, *et al.* Biaxial mechanics and inter-lamellar shearing of stem-cell seeded electrospun angle-ply laminates for annulus fibrosus tissue engineering. *J Orthop Res*, 2013, 6: 864–870.

14. Wan Y, Feng G, Shen FH, *et al.* Biphasic scaffold for annulus fibrosus tissue regeneration. *Biomaterials*, 2008, 6: 643–652.

15. Wu Yaohong, Xu Baoshan, Yang Qiang, *et al.* The physicochemical properties of fibers from long bone tissue engineering scaffold ring and cell biological compatibility. *Chin Orthop J*, 2013, 3: 285–290.

16. Bowles RD, Williams RM, Zipfel WR, *et al.* Self-assembly of aligned tissue-engineered annulus fibrosus and intervertebral disc composite via collagen gel contraction. *Tissue Eng Part A*, 2010, 4: 1339–1348.

17. Gruber HE, Hoelscher G, Ingram JA, *et al.* Culture of human annulus fibrosus cells on polyamide nanofibers: Extracellular matrix production. *Spine (Phila Pa 1976)*, 2009, 1: 4–9.

18. Koepsell L, Zhang L, Neufeld D, *et al.* Electrospun nanofibrous polycaprolactone scaffolds for tissue engineering of annulus fibrosus. *Macromol Biosci*, 2011, 3: 391–399.

19. Nesti LJ, Li WJ, Shanti RM, *et al.* Intervertebral disc tissue engineering using a novel hyaluronic acid-nanofibrous scaffold (HANFS) amalgam. *Tissue Eng Part A*, 2008, 9: 1527–1537.

20. Xu Hai Wei, Xu Baoshan, Yang Qiang, *et al.* Preparation and evaluation of a new type of annulus fibrosus and nucleus pulposus biphasic scaffold. *Chin J Reconstruct Surg*, 2013, 4: 475–479.

21. Hu Y, Zhang Q, Yo R, *et al.* The relationship between secondary structure and biodegradation behavior of silk fibroin scaffolds. *Adv Mat Sci Eng*, 2012, 1: 1–5.

22. Yao J, Turteltaub SR, Ducheyne P. A three-dimensional nonlinear finite element analysis of the mechanical behavior of tissue engineered intervertebral discs under complex loads. *Biomaterials*, 2006, 3: 377–387.

23. Park SH, Gil ES, Mandal BB, *et al.* Annulus fibrosus tissue engineering using lamellar silk scaffolds. *J Tissue Eng Regen Med*, 2012, 3: 24–33.

24. Kim HJ, Kim UJ, Kim HS, *et al.* Bone tissue engineering with premineralized silks scaffolds. *Bone*, 2008, 6: 1226–1234.

25. Zhuang Ying, Chen Jianming, *et al.* Preliminary study on tissue engineering scaffold material and its physical and chemical properties. *Chin J Surg*, 2013, 13: 1334–1339.

26. Laura C, Estelle C, Diego VB, *et al.* Type II collagen-hyaluronan hydrogel — A step towards a scaffold for intervertebral disc tissue engineering. *Eur Cells Mat*, 2010, 20: 134–148.

27. Nosikova Y, Santerre JP, Grynpas MD, *et al.* Annulus fibrosus cells can induce mineralization: An *in vitro* study. *Spine J*, 2013, 4: 443–453.

28. Ganey T, Hutton WC, Moseley T, *et al.* Intervertebral disc repair using adipose tissue-derived stem and regenerative cells. *Spine*, 2009, 21: 2297–2304.

29. Richardson SM, Curran JM, Chen R, *et al.* The differentiation of bone marrow mesenchymal stem cells into chondrocyte-like cells on poly-l-lactic acid (PLLA) scaffolds. *Biomaterials*, 2006, 22: 4069–4078.

30. See EY, Toh SL, Goh JC. Effects of radial compression on a novel simulated intervertebral disc-like assembly using bone marrow-derived mesenchymal stem cell — Cell sheets for annulus fibrosus regeneration. *Spine*, 2011, 21: 1744–1751.

31. Park SH, Gil ES, Cho H, *et al.* Intervertebral disk tissue engineering using biphasic silk composite scaffolds. *Tissue Eng Part A*, 2012, 5–6: 447–458.

32. Bhattacharjee M, Miot S, Gorecka A, *et al.* Oriented lamellar silk fibrous scaffolds to drive cartilage matrix orientation: Towards annulus fibrosus tissue engineering. *Acta Biomaterialia*, 2012, 9: 3313–3325.

33. Bhardwaj N, Kundu SC. Silk fibroin protein and chitosan polyelectrolyte complex porous scaffolds for tissue engineering applications. *Carbohydrate Polymers*, 2011, 2: 325–333.

34. Garcia-Fuentes M, Meinel AJ, Hilbe M, *et al.* Silk fibroin/hyaluronan scaffolds for human mesenchymal stem cell culture in tissue engineering. *Biomaterials*, 2009, 28: 5068–5076.

35. Jao D, Mou X, Hu X. Tissue regeneration: A silk road. *J. Funct. Biomater.*, 2016, 7(3): 22. https://doi.org/10.3390/jfb7030022

# Index